STRUCTURES

STRUCTURES

The Way Things Are Built

Nigel Hawkes

MACMILLAN PUBLISHING COMPANY

New York

Maxwell Macmillan International

New York   Oxford   Singapore   Sydney

First published in the U.S.A. in 1990
by Macmillan Publishing Company, a division of Macmillan, Inc.

A Marshall Edition
STRUCTURES
was conceived, edited, and designed by
Marshall Editions
170 Piccadilly
London W1V 9DD

Macmillan Publishing Company
866 Third Avenue, New York, NY 10022

Collier Macmillan Canada, Inc.
1200 Eglinton Avenue East, Suite 200
Don Mills, Ontario M3C 3N1

Library of Congress Cataloging-in-Publication Data
Hawkes, Nigel.
        Structures: The way things are built/
        Nigel Hawkes.—Marshall ed.
            p.   cm.
        Includes bibliographical references.
        ISBN 0-02-549105-9
        1. Civil engineering—History.   2. Buildings—History.
        3. Underground construction—History.   I. Title.
TA19.H28   1990   90-35928   CIP
624'.09—dc20

Macmillan books are available at special discounts for bulk purchases for sales
promotions, premiums, fund-raising, or educational use. For details contact:

Special Sales Director
Macmillan Publishing Company
866 Third Avenue
New York, NY 10022

10 9 8 7 6 5 4 3 2 1

Originated by Reprocolor Llovet SA, Barcelona, Spain
Typeset by Servis Filmsetting Limited, Manchester, UK
Printed and Bound in Spain by Printer Industria Grafica SA, Barcelona

Managing Editor: Ruth Binney
Editor: Anthony Lambert
Editorial & Picture Research: Elizabeth Loving
Art Director: John Bigg
Art Editor: Peter Laws
Gazetteer: Gwen Rigby and Anthony Lambert
Production: Barry Baker
                Janice Storr

The publisher and author would like to thank Dr Alfred Price for his assistance with
'Factory beneath a Mountain'.

Contents

Introduction

hat has the Panama Canal in common with the Vatican Palace, or Mount Rushmore with the Great Wall of China? All are striking and singular human achievements, expressions of that streak of megalomania that drives the world's great builders. The urge to create something big and memorable—to leave a deep footprint in history—seems to be common to people of every culture and historical period. Behind every great structure there is usually a great man (or, less often, a woman): an engineer, an architect, a priest, a war-lord, a king or a president.

The oldest structures in this book are those erected by the Egyptians, while the newest are the huge scientific instruments in Chile and Switzerland designed to study the infinities of space and the infinitesimal structure of the atom. Between the two lies the whole history of building and civil engineering: a collection of palaces, churches, sculptures, bridges, dams, canals, railways and tunnels which qualify for inclusion by their boldness, scale, or sheer eccentricity.

The book does not try to be exhaustive, nor does it follow absolute rules. Most of the structures described are the biggest or the most important of their kind, but this principle has not been followed slavishly. Some are here because they tell an interesting story or represent a landmark in technique, others because their curious qualities commend them. Some equally worthy structures are excluded on the grounds of familiarity: many books have been written about the pyramids of Egypt, but the much larger pyramid of Cholula in Mexico remains obscure, largely unvisited and poorly understood. Interesting or important buildings denied a full description for lack of space are listed in the gazetteer.

The Location of the Structures

The earliest structures described in this book are mostly located in the Middle East, the cradle of Western civilization, and are associated with the religious life of their times. Religion continues to inspire many remarkable buildings, but the Industrial Revolution gave to Europe and North America a head start in new fields of secular construction achievements. New ways to produce metals in huge quantities, and the invention of new building materials such as reinforced concrete, enabled the Western world to push back the limits of building technology with unprecedented rapidity. This quest for improved materials has continued into the field of space exploration.

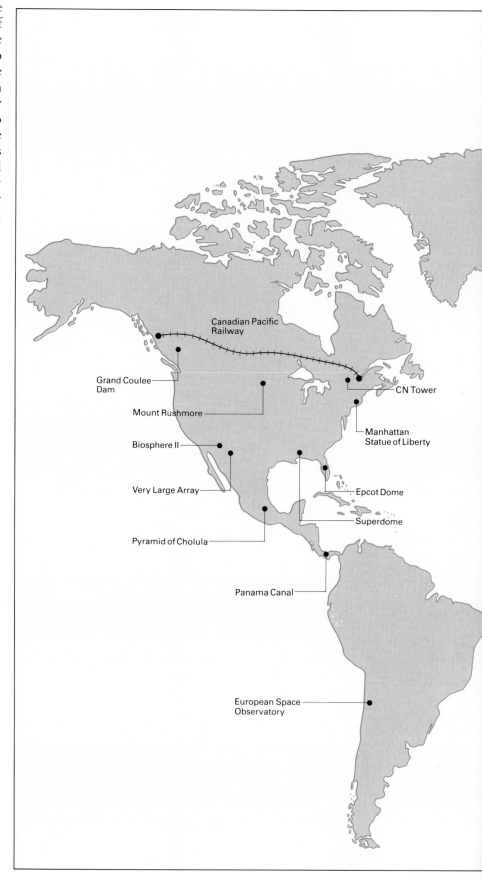

Canadian Pacific Railway

Grand Coulee Dam

CN Tower

Mount Rushmore

Manhattan
Statue of Liberty

Biosphere II

Very Large Array

Epcot Dome

Superdome

Pyramid of Cholula

Panama Canal

Space
Telescope

European Space
Observatory

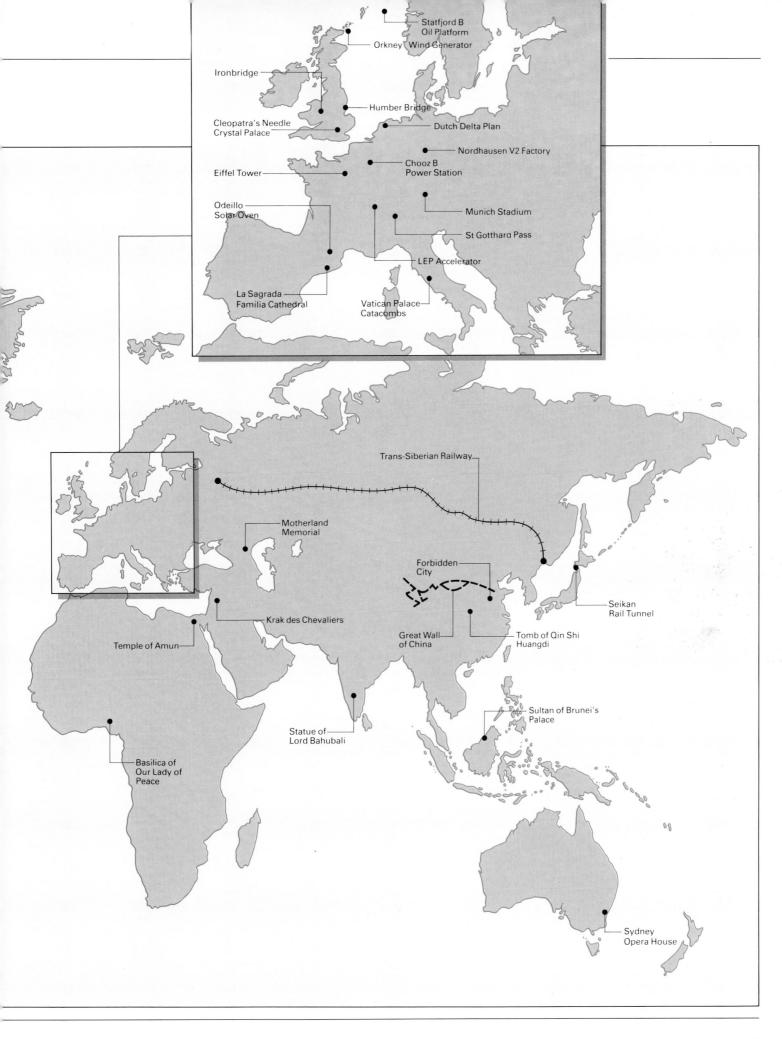

Statfjord B
Oil Platform

Orkney Wind Generator

Ironbridge

Humber Bridge

Cleopatra's Needle
Crystal Palace

Dutch Delta Plan

Nordhausen V2 Factory

Eiffel Tower

Chooz B
Power Station

Odeillo
Solar Oven

Munich Stadium

St Gotthard Pass

LEP Accelerator

La Sagrada
Familia Cathedral

Vatican Palace
Catacombs

Trans-Siberian Railway

Motherland
Memorial

Forbidden
City

Seikan
Rail Tunnel

Krak des Chevaliers

Great Wall
of China

Tomb of Qin Shi
Huangdi

Temple of Amun

Statue of
Lord Bahubali

Sultan of Brunei's
Palace

Basilica of
Our Lady of
Peace

Sydney
Opera House

CHAPTER ONE

Monolithic Memorials

Sculptors know that to achieve an impact, ostensibly life-size statues gain by being rather larger than life. Praxiteles made his statue of Hermes carrying the infant Dionysus 7 feet tall, while Churchill's statue in Parliament Square, London, overtops the living Churchill by quite a margin. Michelangelo went even further: his David was more than 14 feet tall because it was originally intended to occupy a space high on the Cathedral of Florence. But few sculptors have had the daring, the equipment, the time — or the money — to take this process to its logical conclusion.

Huge, monumental sculptures many times life size have the ability to strike directly at the emotions — or so the sculptor of Mount Rushmore, Gutzon Borglum, believed. In an age when everything was big, sculpture, he asserted, needed to be gigantic. It was not a new idea. The Egyptians, with the Sphinx and their obelisks carved from a single block of stone, had already proved that sheer size can take the breath away. In the tenth century AD, Jaina sculptors in southern India produced from the solid rock a holy image of Lord Bahubali so vast that tourists still come to marvel at it today. And Borglum himself had been involved in an unsuccessful attempt to improve the best-known monumental sculpture of all, the Statue of Liberty in New York Harbour.

The first Chinese emperor, Qin Shi Huangdi, opted for numbers rather than size, surrounding his mausoleum with no less than 8,000 warriors made of fired clay. The effect was the same: the terracotta army was meant to impress by its sheer bulk, and it does. So does the gigantic Motherland statue in Volgograd, even though as a work of art it is not a great success.

Huge sculptures have no purpose but to add glory to an image. They celebrate a victory, an idea, or a human life. They speak powerfully across the centuries and the cultures in a language that all can understand, enduring when smaller works of art are lost, transported, or destroyed. Of all structures, these are the ones most likely to survive into eternity.

Monolithic Memorials
Cleopatra's Needles
Tomb of Qin Shi Huangdi
Statue of Lord Bahubali
Statue of Liberty
Mount Rushmore
Motherland Memorial

London's Egyptian Obelisk

Fact file

Obelisks 3,500 years old that are a tribute to the craftsmen of ancient Egypt

Creator: Tuthmosis III

Built: c1504–1450 BC

Material: Granite

Height: London 68ft 6in; New York 69ft 6in

Weight: London 186 tons; New York 200 tons

Among the most remarkable memorials of the ancient civilization of Egypt are obelisks, slender pillars with four flat sides tapering to a point, cut from a single piece of granite. Highly polished and decorated with inscriptions and drawings, the obelisks were created almost 4,000 years ago with the most basic of tools.

The largest, weighing 455 tons and more than 105 feet high, was commissioned by the Pharaoh Tuthmosis III and now stands in the Piazza San Giovanni in Laterano, Rome, but an even larger one still lies unfinished in the quarry near Aswan from which the obelisks came. Two of the most interesting, also commissioned by Tuthmosis III, were erected like sentinels at the entrance to the Temple of the Sun at Heliopolis, just north of modern Cairo. Later these two acquired the name of Cleopatra's Needles, and were moved to new sites in London and New York. How were these huge blocks of stone quarried and cut without the use of metals? How were they moved without the use of wheels, and raised upright without cranes, scaffolding or even pulleys?

The purpose of the obelisks appears to have been part religious, part ceremonial. They were erected in honour of the sun god, and the first of them went up at Heliopolis, the main centre of worship. But the inscriptions on the obelisks glorified the achievements of earthly rulers: territories conquered, rivals defeated, and the anniversaries of a king's reign. Cleopatra's Needles carry inscriptions down the centre of each side extolling the virtues of Tuthmosis III, and further hieroglyphics, added 200 years later, record the victories of another great Pharaoh, Rameses II.

Cleopatra's Needles are made from red granite and may have come from the very same quarry in Aswan where the even larger unfinished obelisk still lies. If the quarry workers had not come across an unexpected crack in the granite, forcing them to abandon work, it would have been the biggest obelisk of all, standing more than 135 feet high and weighing 1,168 tons. For archeologists it is among the most precious of finds, for it shows how all the other obelisks were made.

First, the quarry engineers had to locate an area of perfect stone from which an obelisk could be cut in a single piece. This was done by sinking test shafts down into the rock. Once a site had been chosen, the first stage was to smooth the upper surface of the rock by removing uneven areas. Bricks were heated and placed on the surface, then doused with cold water. The effect was to fracture the rock surface, making it easier to remove.

The next stage was to cut down on either side of the obelisk, creating trenches. The clue as to how this was done comes from the discovery of balls of the mineral dolerite lying around the quarry. Such balls, between 4 and 12 inches in diameter, and weighing 10 lb or more, occur naturally in the Eastern Desert, from which they had been brought. Mounted on rammers, the stones were lifted and then brought down with great force on the rock, so crushing it. At any time, several thousand men would have been at work, in groups of three, two standing and raising the ramrod, the other squatting down to direct the blow. To maintain the rhythm, a chanter would intone.

Progress must have been very slow, and it probably took six months to a year for the teams to cut down to the full depth needed. The next step was to detach the bottom of the obelisk from the rock, and according to the English archeologist Reginald Engelbach this was also done by pounding. First an area beneath the obelisk would have to be prepared, to form a gallery in which to work. Then wooden beams would be used to support the obelisk as it was progressively cut away by pounding horizontally. Some believe that wooden wedges were used, either by progressively driving them in to force it to crack, or by inserting the wedges and wetting them, allowing the expansion of the wood to exert the force. Others believe the whole job was done with the dolerite balls.

Decoration of the obelisks may have begun while they were still at the quarry, although the chances are that final decoration, including covering the upper surfaces and the point—the pyramidon—with gold, was left until the obelisk was upright in its final position.

The next problem was getting the obelisk out of its pit in the quarry, on to barges and down the Nile to its final resting place. Hundreds of men would have used great balks of timber as levers, lifting first one side of the obelisk and then the other, packing fresh material under it at every lift. In this way it might have been raised by stages to a height almost equal to that of the surrounding rock, and a passage cut to enable it to be pulled away.

Some doubt exists over whether rollers were used to ease the movement of the obelisks. None has ever been found, but without them up to

6,000 men, pulling 40 ropes, would have been needed to overcome friction. At the edge of the Nile, it is assumed that barges were drawn up and virtually buried in sand, perhaps at a time when the river was low and the barges could rest on the bottom. Then the huge obelisk would be dragged on top of the sand embankments and the sand removed from around and beneath it, allowing it gradually to settle on to the barge. When the river rose at the annual flood, the journey to the obelisk's final destination began.

Tuthmosis III created at least seven obelisks, five in Thebes and two in Heliopolis. Four survive, but not one remains in its original position. The two Cleopatra's Needles have had an extraordinary history. For 1,500 years they stood at Heliopolis, while Egypt fell successively under the rule of the Ethiopians, the Persians, and the Greeks under Alexander the Great. He founded Alexandria, where later Queen Cleopatra ruled as the last of the Ptolemaic dynasty and built a palace on the edge of the Mediterranean dedicated to Julius Caesar.

When Cleopatra died in 30 BC, Egypt fell once

London's Egyptian Obelisk

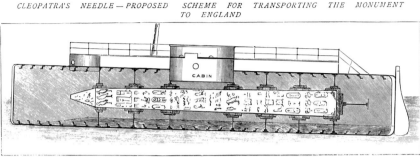

LONGITUDINAL SECTION OF THE CYLINDER

EXTERIOR OF THE CYLINDER CONTAINING THE NEEDLE

more under foreign domination, this time the Roman Empire, and the two obelisks from Heliopolis were moved to a new site at the water gate to Cleopatra's palace. Centuries later they had acquired the name Cleopatra's Needles, although in reality they dated from a period 15 centuries before she was even born.

There they stood for a further 1,500 years as Cleopatra's palace fell into ruin and disappeared. At some point—it is not clear when—one of the two obelisks fell to the ground, where it was seen half-covered in sand by the traveller George Sandys in 1610. When Napoleon Bonaparte landed in Egypt in 1798, with the intention of taking the Middle East from the Turks, it was the Royal Navy and a British army that defeated him. In gratitude, the Turks, restored to power, were glad to accept the suggestion that the British should take home with them the obelisk that had already toppled. Napoleon had had it in mind to take them both, and wires had already been attached to the upright obelisk ready to pull it down.

It was, however, another 65 years before anything was done about this offer, years in which the obelisk lay on the ground and was abused by passing tourists, who were inclined to chip pieces off it as souvenirs. By 1867, the obelisk was in dire peril, for a Greek merchant named Giovanni Demetrio had bought the land on which it lay and was proposing to develop it. Unable to move the obelisk intact, he was ready to break it up and use it as building material. The man who came to the rescue was General Sir James Alexander, who heard of the threat, roused public opinion and helped to devise a scheme for bringing the obelisk to London.

A special tube-shaped craft was designed to carry the obelisk, estimated to weigh 185 tons, on its ocean voyage to England. *Cleopatra*, as the floating cigar-tube was called, was put in tow of a tug, the *Olga*. So long as the weather was calm, all went well, although communication between the two vessels was difficult and the *Cleopatra* pitched like a see-saw. Worse things were to come, however. In the Bay of Biscay, a storm struck, and the tow-rope had to be cut. When the weather eased and *Olga* went in search of the *Cleopatra*, she had completely disappeared.

But *Cleopatra* had not sunk. Lying low in the water, with the seas pouring over her, this curious vessel was spotted by another British ship, the *Fitzmaurice*, which with enormous difficulty towed her into Ferrol harbour upside-

down. From here she was, in due course, recovered, after the salvage claim with the owners of *Fitzmaurice* had seen settled. She was brought into the Thames and her precious cargo raised into its present position on the Embankment. Her twin was transported to the US in 1880 aboard a rather more seaworthy vessel, and erected in Central Park, New York.

The two obelisks—and others, now in Paris, Istanbul and Rome—were taken from Egypt before the modern world had developed a conscience about stripping a nation of its cultural treasures. Rome acquired its collection of 13 obelisks in ancient times, as did Istanbul, while London, Paris and New York obtained theirs in the nineteenth century. The irony is that they all now stand virtually ignored, dwarfed by modern buildings and surrounded by traffic. People who would pay large sums of money to travel to Egypt to see them in their original surroundings never give them a second thought where they stand today.

The tube in which *Cleopatra's Needle was brought to London was built in Alexandria and holed by a sharp stone as soon as it was launched. Once patched and refloated, the cylinder was fitted with 2 keels, a cabin and deck. Higher rates of pay had to be paid by the English captain to the 6 crew, such was the fear about the vessel's safety.*

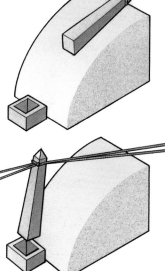

The method the Egyptians used for raising obelisks (above) entailed, according to the French archeologist Henri Chevrier, the construction of huge embankments leading to a curve descending to the plinth. Sand to stop the obelisk falling too fast would be removed, allowing it to settle in a notch in the plinth at an angle of about 34 degrees. Ropes would be used to pull it upright.

Raising of the needle on the Adelphi Steps beside the Thames in September 1878. It was unloaded and hauled up the steps using hydraulic jacks and screw traversers. Once on its plinth, a huge wooden framework was built over the obelisk to raise it sufficiently high to swing it into the vertical plane with steel cables. The needle was so finely balanced about its centre of gravity that the beam holding it could be swung by one man.

The Terracotta Army

Fact file

One of the world's most spectacular burial sites

Creator: Qin Shi Huangdi

Built: 246–209 BC

Material: Terracotta

Number of figures: Approximately 8,000

The soldiers and horses (opposite) have been left in the pits where they were discovered, although extensive restoration has been necessary on the most badly broken figures. It is estimated that about 600 horses, 7,000 warriors and 100 chariots will be unearthed. The Qin horses were bred with powerful lungs to gallop for long distances at high speed.

In March 1974 workers at the Yanzhai Commune, 18 miles from the ancient Chinese capital of Xi'an, were worried that their harvest would be lost as a result of drought. Searching for water, they dug a well and stumbled instead across one of the most spectacular archeological discoveries of the twentieth century. They found a few fragments of terracotta, sculpted and fired into the shape of warriors and horses.

Since then, archeologists have unearthed an entire army of terracotta warriors and believe that ultimately there may be as many as 8,000 of them, figures a little larger than life-size, carved with enormous skill and buried for more than 2,000 years. They provide a glimpse into the world of the first emperor to unite China, Qin Shi Huangdi, a remarkable man who created the first totalitarian society on earth and ruled it with a combination of efficiency and utter ruthlessness.

The warriors were created as an entourage to guard the emperor and to guide him into the next world. He was born in 259 BC, and came to the throne of the state of Qin as a 13-year-old boy in 246 BC. Almost immediately, despite his youth, work began on a splendid tomb which was to be his home after death. It was another 36 years before he occupied it, years in which he annexed the six other independent kingdoms of China and became the first emperor. As a warrior and administrator, Qin Shi Huangdi has had few equals in history. He unified the Great Wall of China, linking together separate walls built by earlier northern states. His army was equipped with swords and arrowheads made of bronze, and crossbows powerful enough to penetrate armour, yet light enough to be carried by mounted archers. The triggers were far more sophisticated than any that were to appear in Europe for many centuries.

Qin Shi Huangdi created a centralized, auto-cratic state, with a uniform code of law, a single currency and set of weights and measures and a written language. He built a network of tree-lined roads, 50 paces wide, radiating outward from the Qin capital, Xianyang. He ruled by force, and fear: the law provided for whole families to be executed for the crimes of one member, and millions of men were drafted into the army and civilian labour force. He permitted no independent thought, burning books and burying scholars alive. He set the pattern of authoritarian rule which has survived in China to this day.

During his life Qin Shi Huangdi created several palaces, plus a huge mausoleum, which has yet to be excavated. The history books tell us that beneath this mound of earth some 250 feet high is a tomb chamber whose ceiling is decorated with pearls, representing the stars, while the stone floor is a map of the Qin empire, the rivers glittering with mercury to represent water. The tomb was filled with treasures and fitted with booby traps in the form of crossbows primed to fire at any intruder. Here the emperor was buried in 209 BC, the year after his death. Buried alive with him were his wives—none of whom had borne him any children—and the craftsmen who knew the secrets of the tomb.

Whether these stories are true must await the excavation of the tomb. The terracotta army, clearly designed as a guard, was found about a mile east of the mausoleum, and all of its number face eastward, perhaps because the emperor expected any revengeful attack from the six conquered kingdoms to come from that direction. Scholars have speculated that this was merely a storage place—but why the warriors did not join the emperor after his death is not clear. As guards, they proved ineffective. Within three years of his death, the emperor's tomb had been plundered by a rebel general, Hsiang Yu, who also discovered the terracotta army in its underground vaults. He ordered the roof to be set alight and it collapsed on the figures, breaking many of them and covering them in mud.

The warriors have been found in three separate pits. The largest contains up to 6,000 soldiers and more than 100 horses. It is more than 250 yards long, 70 yards wide, and 17 feet deep. The floors are paved with brick and the pit consists of a series of trenches or corridors, divided by earthen walls, and roofed originally with beams, woven mats, and alternate layers of plaster and earth up to ground level. So far, about 1,000

The Terracotta Army

warriors and 24 horses have been excavated, a small fraction of the total the pit is believed to contain. The horses, in teams of four, pull wooden chariots of which little has survived the centuries. The other two pits are similar, but smaller, with about 1,000 warriors in the second and 68 in the third. From the arrangement of the third pit, it appears to have represented the command headquarters where officers controlling the other two pits were stationed.

The warriors themselves range in height between 5 feet 8 inches and 6 feet 5 inches, somewhat larger than the average height in Qin times. They were made by a combination of moulding and hand-modelling. Several dozen different moulds were used to make the heads, producing rough castings which were then finished off by hand to create individuality.

For each warrior, ears and moustaches were made in separate moulds and attached later, and headgear, lips and eyes also show signs of having been made separately. The material of which they were made is clay, which when fired shrinks by about 18 percent, so the unfired figures must have been considerably larger. Heads and bodies were made separately, then joined together. To ensure that they stood upright, thicker layers of clay were used in the lower parts of the body.

The faces of the warriors have been classified by some scholars into 30 different types, but fall into ten broad categories, described by the Chinese written characters they most nearly resemble. The face like the character which is pronounced "you", for example, belongs to the mightiest of the warriors, and is broader in the cheekbone than in the forehead. The opposite face, with forehead broader than cheekbones, represents the Chinese character "jia" and is found most commonly among the vanguard, since its appearance is alert and resourceful. Many of the faces have tightly closed lips and staring eyes, to create the feeling of bravery and steadiness. Others show vigour, confidence, thoughtfulness, or experience.

The skill of the carvers has created an army of recognizable types, no two of whom are exactly the same. They are neither mechanical copies of real warriors, nor mere imaginative figures; rather, they represent a gallery of ideal types such as one might find in a well-run army, ranging from the young and enthusiastic subaltern to the wise and experienced sergeant.

The armour worn by the warriors is also skilfully executed, showing that the sculptors

were familiar with Qin armour. The pieces of battledress were individually made, and fit their wearers perfectly. The horses are carved with similar vigour, and meet the requirements for fine horses laid down by Qin writers several hundred years earlier: pillarlike forelegs, bowlike hind legs, high hoofs, slim ankles, flared nostrils and a broad mouth. The horses' saddles are decorated with tassels, and were originally painted red, white, brown and blue and made to represent leather. There are no stirrups, suggesting that the horsemen in the emperor's army had no need of them.

In December 1980 a remarkable find was made 20 yards to the west of the emperor's tomb mound: a pair of bronze chariots, complete with horses and charioteers. Unlike the terracotta warriors, these are two-thirds life-size, but are modelled with even greater delicacy and skill.

The hangar protecting the pits covers an area of 19,000 square yards. The burial of the army replaced the earlier practice of human and animal sacrifice. The army was deployed with archers at the front, chariots on the right flank and the cavalry on the left, surrounding lines of infantry interspersed with chariots.

The brick-paved floors can be seen beneath these warriors (left), who all face east. It is thought the direction may have reflected Qin Shi Huangdi's fear that a revengeful attack would be most likely to come from the 6 kingdoms he conquered. The depth of the figures indicates why the army was found only when farmers were drilling for water.

The armour of the warriors (above right) has provided invaluable information on the state of military technology; this is an infantryman. It is now known how armour was designed for generals, cavalrymen and charioteers. Even the hair of warriors (right) was individually finished, as well as all their facial features.

The bronze chariots have survived far better than the wooden ones pulled by the terracotta horses, and provide a reliable model of what a Qin chariot was actually like. Horses and charioteers were originally painted, but have now faded to a greyish white. The horses' bridles are decorated with gold and bronze decorations. Few doubt that other remarkable finds will be made as excavation of the tomb proceeds.

The entire army, evidence of the supreme power and megalomania of Qin Shi Huangdi, must have taken hundreds of craftsmen many years to complete. If it was supposed to preserve the emperor in death, it failed; but it has brought down to us an extraordinary insight into the world of China's first emperor, a world in which supreme craftsmanship coexisted with cruelty and violence. No more spectacular memorial exists than Qin Shi Huangdi's terracotta army.

How the warriors may have looked

The Museum near the excavated pits contains a display of figures painted as it is thought they once looked. Firing the fine-textured clay figures gave them a smooth finish which was then painted with pigments mixed with gelatine to give them an even more life-like appearance. Only traces of this paint remain.

The Granite Icon

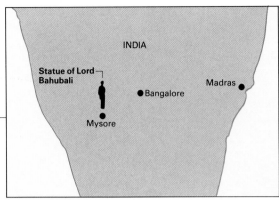

Fact file

Believed to be the tallest monolithic statue carved out of a single piece of rock

Builder: King Chamundaraya

Built: 981

Material: Granite

Height: 57 feet

Every dozen years or so, followers of the Jaina religion make a pilgrimage to a small town, Shravana Belgola, in Karnataka in southern India. There, on top of a hill, stands a monolithic statue of a naked man, some 57 feet high and carved out of solid granite a thousand years ago. From a platform erected around the head of the statue, the pilgrims pour water, milk, ghee, curds and sandalwood paste in various colours over the holy image in a ceremony as ancient as the statue itself. The great statue gleams in the sun as the priests chant mantras and sound gongs. The ceremony was last performed in 1981, the thousandth anniversary of the statue's creation.

The story of how this remarkable statue was carved is lost in the mists of time, and buried by the accretion of myths and legends which surround it. The statue depicts Lord Bahubali, one of the sons of Rishabha, the founder of the Jaina religion and a king who decided to renounce worldly power in the search for righteous living and salvation. Realizing the ephemeral nature of success in this world, Rishabha abandoned his two wives and more than 100 children to seek enlightenment in the forest. Before leaving he appointed one son, Bharata, to run his kingdom of Ayodhya, and awarded another, Bahubali, the principality of Pondnapura. Bharata became a powerful ruler, but of his brothers only Bahubali refused to accept his overlordship.

The two brothers fought, according to legend, first by staring one another down, then by fighting in water, and finally by wrestling. Bahubali won all three contests, and at the end of the wrestling bout held his brother high and prepared to dash him to the ground. Suddenly he was overcome by remorse and disillusionment, and put Bharata gently down. Without a moment's hesitation, he then left for the forests, pulled out his hair and stood still, his arms at his sides and his feet straight ahead, as he sought enlightenment. He stood like this for a whole year, while ants built nests around his feet and the creepers of the forest began to wind around

The climb to the statue up 614 steps carved out of the rock begins near a "tank" in Shravana Belgola (left). Since the statue is concealed by cloisters built around it during the 12th century, ascending pilgrims see it in its totality only at the last moment. The dramatic impact of losing sight of the statue and then suddenly seeing it again at close quarters (right) is awe-inspiring. The statue, here with scaffolding ready for the annointing ceremony, is visible from 15 miles away.

The Granite Icon

his legs. Finally Bharata and two of Bahubali's sisters came to the forest to offer homage, and his resentment and pride finally disappeared. Bahubali then achieved the highest states of enlightenment known to the Jaina religion.

Bahubali's statue, on top of a hill 3,350 feet high, is believed to have been created in AD 981 under the command of a powerful general named Chamundaraya, a king of the Ganga dynasty. Contemporary descriptions of how the work was done do not exist—a long inscription on an elaborately decorated column erected at the same time by Chamundaraya, which might have held the secret, was obliterated when a later record was inscribed on it in 1200.

At the site of the statue itself, inscriptions in three languages—Kannada, Tamil and Marathi—make it clear that it was Chamundaraya who had the image made, but do not say how it was done. Since the statue is enormous and carved from a single piece of granite, it must have taken many workmen years to create it.

The statue is carved in the round from the head to the lower half of the thighs, with the rest in bold relief. The shoulders are broad, the waist narrow, and from the knees downward the legs are somewhat out of proportion. The arms hang straight down the sides with the thumbs turned outward. The whole figure exudes serenity. At Lord Bahubali's feet are carved anthills with serpents, and creepers wind their way up his legs. The granite of which the statue is carved is smooth, homogenous and hard—the ideal material for such a huge work of art. Ever since its creation the image has been one of the wonders of India.

The statue has many names and several curious features. It is often called the Gommata or Gommateshwara, either because Gommata was another name for Chamundaraya, because the word means beautiful, or because it means a hill or hillock. Its only visible flaw is a shortened forefinger on the left hand, and a variety of explanations has been put forward to explain this. One is that Chamundaraya ordered the finger to be mutilated because the image, when the carvers had finished their work, was simply too perfect. By deliberately damaging it, he sought to avert the evil eye.

Another theory is that the statue was damaged as an act of revenge in the reign of King Vishnuvardhana in the twelfth century. The king, who had lost a finger, was angered when a Jaina guru refused to accept food from his

The head-annointing ceremony, or Mahamastakabhisheka, *is known to have taken place in 1398 and has since been held regularly. Women are responsible for pouring offerings over the head of the statue. The influx of a million pilgrims (right) for the ceremony in 1981 necessitated construction of 7 satellite towns.*

The naked Indras, priests of the Jaina temples, direct the ceremony of Mahamastakabhisheka. *In 1780 it was recorded that they worshipped the 1,008 shining metal pots that were used to carry sacred water up the hill to be poured over the head of Bahubali.*

mutilated hand. In revenge, he abandoned the Jaina religion and ordered the statue damaged as an act of revenge. After such colourful legends, it is almost a disappointment to record that the most likely reason for the flaw is that the carvers came across a fault in the rock, which caused the end of the finger to fall off. To make the best if it, they carved a fingernail on to the end of the shortened finger.

The conservation of the statue of Lord Bahubali is the responsibility of the Archaeological Survey of India. Since it has stood unprotected for more than a millenium, time has begun to weather its smooth grey surface. In addition, the fact that huge quantities of milk, ghee and curds have been poured over the statue at irregular

Tall, ornate scaffolding is required to enable the offerings to be poured over the head of Lord Bahubali. Once the statue was finished, King Chamundaraya arranged to have it annointed with milk. But however much he poured over its head, the milk would not descend below the naval. Then an old woman named Gullakayajji arrived with a few drops of milk in the skin of an eggplant, and poured it over the statue. It not only covered the statue but flowed down to the valley, forming a pond. This miraculous event is commemorated with a statue to Gullakayajji, within the cloister that surrounds the statue.

intervals has resulted in an accretion of grease, and the growth of moss and lichen. More alarming were the appearance of small cracks all over, and particularly on the face, and the development of pitted areas where the stone had begun to flake off. Beginning in the early 1950s, a series of experiments was carried out to determine the best way of removing the grease, cleaning and repairing the statue.

Before the ceremony of *Mahamastakabhisheka*, the statue now receives a preliminary coat of paraffin wax in solvent oil, which allows the greasy materials in the milk and other offerings to flow soothly over the stone, without getting into its pores. Such treatment also makes it much easier to clean the statue afterwards.

Although it is the largest, the statue of Lord Bahubali at Shravana Belagola is not the only iconographic image of him. There are four copies of it, the biggest at Karkal, which was carved in 1432 and stands around 40 feet tall. A second version at Enur, carved in 1604, is 33 feet tall. A magnificent bronze Bahubali, dating from the ninth century, is in the Prince of Wales Museum of Western India in Bombay. But none can compete with the original, its grandeur enhanced by the mystery of its creation. It stands, as one writer put it, "like a giant over the rampart of an enchanted castle, uninjured, though darkened by the monsoons of centuries, its calm gaze directed eastward toward a nearby mountain range mantled with forests."

Offerings at the feet of the statue indicate the the variety of substances with which its head is annointed: coconut milk, yoghurt, ghee, bananas, jaggery, dates, almonds, poppy seeds, milk, gold coins, saffron, yellow and red sandalwood pastes, sugar and, at the ceremony in 1887, precious gems of 9 kinds.

Monument to Freedom

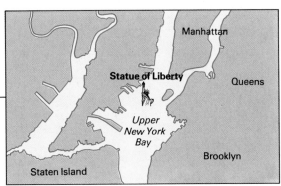

Fact file

The tallest monument in the world when given by France to the United States of America to mark the centenary of its independence from Great Britain

Designer: Frédéric-Auguste Bartholdi

Built: 1875–86

Materials: Copper, iron

Height: 306 feet 8 inches

Bedloe's Island, a 12-acre island in Upper New York Bay, provided the perfect site for the statue. Named after its owner, Isaac Bedloe, the island is visible to all shipping using the harbour and entering the Hudson River, making the statue a prominent landmark.

No statue expresses a more powerful symbolism than the colossal sculpture of a woman holding aloft the torch of freedom which dominates the harbour of New York. For 17 million immigrants from Europe, for whom she was the first sight as they reached American shores, the Statue of Liberty meant a new life in a new land. One of those immigrants remembers: "She was a beautiful sight after a miserable crossing that September. She held such promise for us all with her arm flung high, the torch lighting the way."

This was exactly how the statue's sculptor, Frédéric-Auguste Bartholdi, had envisaged his great lady: "Grand as the idea which it embodies, radiant upon the two worlds". The suggestion of such a statue was first made at a dinner party held near Versailles in 1865 by the historian and politician Edouard de Laboulaye. It was intended to symbolize the friendship between France and the United States at the time of the American Revolution, and mark the hundredth anniversary of the US as a nation. Bartholdi, a young sculptor with an established reputation, was a guest at the dinner and supported the idea. When he visited the US in 1871 he quickly identified the right site, on a small island in Upper New York Bay, south west of Manhattan, and back in France he began work on the first small models of a woman holding a torch.

Raising the money for the Statue of Liberty did not prove easy. Lotteries and dinner parties held by the Franco-American Union in France eventually financed the statue, while the huge pedestal upon which it was to stand was built from American contributions, raised with the powerful support of Joseph Pulitzer and his newspaper, *The World*. Bartholdi settled upon a statue built of beaten sheets of copper, mounted on a framework of iron: bronze or stone would have been too expensive and too heavy to transport. He knew the method would work, for he had seen the seventeenth-century statue of St

Carlo Borromeo, by G.B. Crespi, at Lake Maggiore in Italy, which stands 76 feet high. Bartholdi decided to create a statue twice as high, the largest in the world.

To design the supporting structure he first consulted Eugène-Emmanuel Viollet-le-Duc, the high priest of the Gothic revival in France. But he died in 1879 with the task incomplete, and Bartholdi turned to Gustave Eiffel, a daring engineer and specialist in structures of iron. Eiffel proposed supporting the statue on a central iron tower anchored firmly in the pedestal. The tower would consist of an iron truss, with diagonal bracing. From this strong framework, a secondary structure would be hung, approximating to the shape of the statue, and from this secondary framing a series of flat, springy iron bars would connect directly to the statue's skin.

The skin of the statue consists of 300 copper plates, beaten into shape by the technique known as *repoussé*. First Bartholdi made a series of clay models, of increasing size, in which he refined and perfected the form of the statue. From Bartholdi's one-third scale models, craftsmen at the workshops of Gaget, Gauthier et Cie in Paris made full-scale plaster copies, which were then used to form moulds by surrounding them with a wooden framework. The copper sheets were then created by beating them into shape on the inside of the wooden moulds. Thin copper sheets only $\frac{3}{32}$ of an inch thick were used, each overlapping with the next, and rivet holes were drilled to fix neighbouring sheets together. To ensure that the method really worked, the statue was temporarily assembled in the courtyard of Gaget, Gauthier. By 1885—nine years late for the bicentennial it was supposed to celebrate—the statue was finally on its way to New York.

There, work on the huge pedestal had also been delayed. It had been designed by the American architect Richard Morris Hunt, a specialist in the *beaux-arts* style. The pedestal is no mean construction in itself, standing 89 feet high on a 65-foot foundation. The style eventually selected by Hunt is vaguely Egyptian in style, strong and simple, and enhances the statue placed on top of it. Building of the pedestal began in 1883 and was completed in 1886, by which time the sculpture had been waiting for 15 months in its crates. It was removed and reassembled, working upward from the ground without exterior scaffolding. As the framework

Monument to Freedom

rose, the workers attached the skin, leaning over the side to install the rivets. In October 1886 the statue was finally dedicated.

The methods of construction worked well, but had one important flaw. The iron armature rods reacted electrolytically with the copper skin, causing corrosion which, by 1980, had destroyed much of the strength of the structure. The rods had swollen, causing rivets to shear or fall off and allowing rainwater to seep in, which exacerbated the problem. The inner surface of the copper had been painted many times, in an attempt to preserve it, but that had trapped water and in some places it was only the paint that was keeping together pieces of broken armature bar. The torch, and the structure supporting the raised right arm, were in particularly bad shape.

A major restoration project was launched to ensure that the statue would survive a second century. Every armature rod in the structure was replaced with new rods of stainless steel, working slowly through the statue and replacing a few at a time so that it would retain its integrity. The old bars were removed, and exact copies made in the new material and replaced, using the original rivet holes. It took a year to replace 10,000 feet of armature bars.

The most important repair of all was the replacement of the torch. Bartholdi had wanted to make the torch glow by projecting a strong light from the torch platform on to the flame itself, which was gilded, but the plan was dropped at the last minute for fear that the strong light would dazzle ships' pilots in the harbour. Instead, portholes were cut and lights

Full-size plaster copies of the statue's components were made in Paris, around which was built a "negative" wooden form, reversing the model's contours. Traditional wood lath-and-plaster techniques were used to make the life-size copies.

installed within, producing a feeble result which Bartholdi compared with the light of a glow-worm. In 1916, the American sculptor Gutzon Borglum, creator of Mount Rushmore, converted the flame into a lantern by cutting holes in it, installing amber-coloured glass, and mounting a light inside. The flame was now nothing like the one Bartholdi had designed, and it also began to leak, creating further corrosion which weakened it.

By the time of the 1980s restoration, the flame was in such poor condition that it required complete replacement. The decision was taken to restore the lamp to as near Bartholdi's original as possible, and to remove the Borglum lantern and display it in the statue's museum. Appropriately a French firm, Les Métalliers Champenois from Rheims, won the contract for the job, and produced a gilded flame as close to the original as could be contrived. Modern lights—far more intense than any available for use in Bartholdi's day—were installed, so that today the flame glows at night just as he originally hoped it would.

The incongruous sight of Liberty towering over the Paris factory of Gaget, Gauthier. Attaching the copper sheets from the bottom up, this temporary assembly used only 1 rivet in 10 so that it could be easily taken apart and packed into 210 crates for shipment to New York.

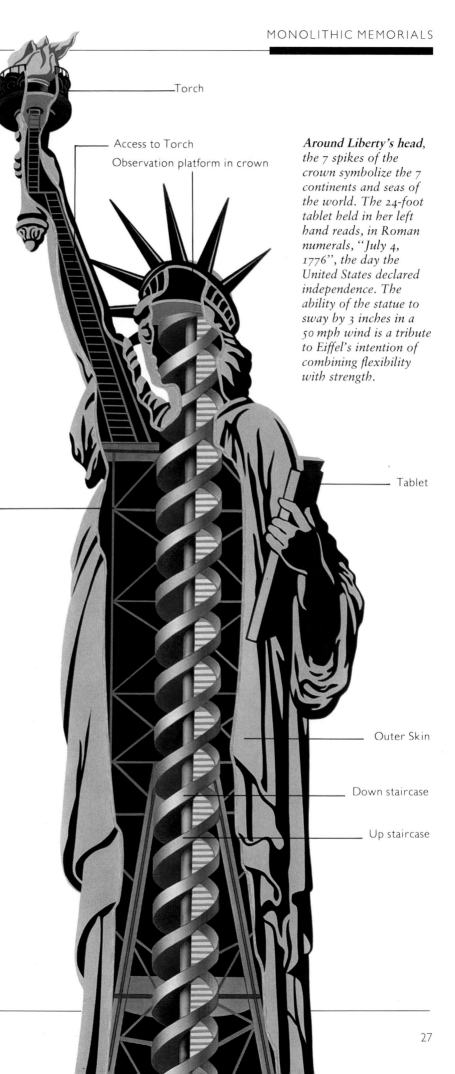

Torch

Access to Torch

Observation platform in crown

Around Liberty's head, *the 7 spikes of the crown symbolize the 7 continents and seas of the world. The 24-foot tablet held in her left hand reads, in Roman numerals, "July 4, 1776", the day the United States declared independence. The ability of the statue to sway by 3 inches in a 50 mph wind is a tribute to Eiffel's intention of combining flexibility with strength.*

Tablet

Outer Skin

Down staircase

Up staircase

The torch and flame *were the first parts of the statue to reach the United States, being sent to the Philadelphia Centennial Exposition in 1876. After being displayed in Madison Square Park, New York, they were returned in 1883 to Paris where the design of the flame was altered.*

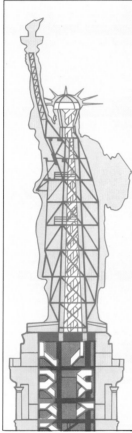

A computer diagram *(left) of the statue's iron skeleton, made during the statue's restoration between 1982 and 1986. The network of armatures attached to the central tower was the only part based on Viollet-le-Duc's ideas. The two spiral stairways (above) connect the base with the crown, which has a 25-window observation platform for 30 visitors.*

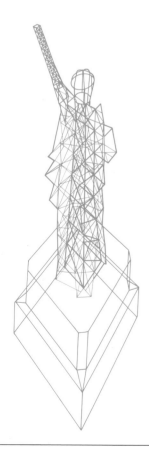

27

Builders of a Nation

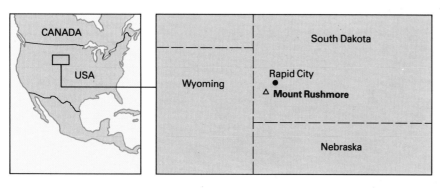

Fact file

The world's largest sculpture, carved out of a granite cliff face

Sculptor: Gutzon Borglum

Built: 1927–41

Material: Granite

Height: 60 feet

Rock excavated: 450,000 tons

Foreman William Tallman hanging from the lower rim of Jefferson's eyelid, when the eye was still at a rough stage. To prevent the eyes having a dead look, Borglum carved a ring, or pupil, several feet across and deep enough to ensure that it was always in shadow; in the centre a peg of rock was left to reflect light within the pupil.

In the Black Hills of South Dakota, USA, the sculptor Gutzon Borglum created the world's most gigantic piece of sculpture by carving the faces of four American presidents into a granite cliff. Each face is about 60 feet high, and some 450,000 tons of rock had to be removed by explosives, pneumatic drills and chisels to create the group. They have noses 20 feet long, mouths 18 feet wide and eyes 11 feet across. Given bodies on the same scale, the four presidents depicted—George Washington, Thomas Jefferson, Abraham Lincoln and Theodore Roosevelt—would stand about 460 feet tall.

The creation of Mount Rushmore took more than 15 years, though most of that time was spent not on the rock but on the stump, trying to raise the money needed. The idea came from Doane Robinson, a lawyer and writer who in the early 1920s held the post of official historian in South Dakota. In 1923 it occurred to him to encourage more tourists to come to the state by commissioning a massive sculpture in the Black Hills. After failing to interest the sculptor Lorado Taft, Robinson put the idea to Gutzon Borglum. He had found the right man, perhaps the only man, with the confidence and skill needed to turn his dream into a reality.

Borglum was a successful sculptor with a taste for publicity and very little tact. The son of a Danish immigrant, he studied art in San Francisco and Paris, and worked for three years in London before establishing himself in New York. He created 100 statues for the Cathedral of St John the Divine in New York, and developed a taste for huge sculpture with a head of Abraham Lincoln carved from a 6-ton block of marble, now in the Capital Rotunda in Washington.

By then he was convinced that huge sculptures were needed to match the spirit of what he called the Colossal Age. "Volume, great mass, has a greater emotional effect upon the observer than quality of form", he wrote. "Quality of form affects the mind; volume shocks the nerve or soul centres and is emotional in its effect." Grumbling that there was not a monument in America as big as a snuff box, Borglum set out to remedy the deficiency. He found his canvas at Mount Rushmore, a great cliff of granite 400 feet high and 500 feet long, towering like a battlement of stone above pine trees and fresh vegetation.

Although Borglum asserted more than once that raising the money for the job would be simple, this was far from the truth. Eventually, however, Borglum, Robinson and the two senators from South Dakota persuaded Congress to set aside $250,000, half the expected cost, with the rest to be raised by public donation. The bill passed Congress and was signed into law by Calvin Coolidge just in time; within months the Stock Market crash of 1929 had wiped out whole fortunes and no money would have been available for such an apparently frivolous enterprise as carving faces on a mountain.

Outwardly, Borglum was certain that now he had sufficient money to make a start he could do the job. But he knew very little about the rock and was ignorant as to whether it would prove to be workable. Nor did he set out with any preconceived ideas of what the finished sculpture would look like. The head of Washington, he decided, must be dominant, so he set out to sculpt it without finally deciding where the other heads would ultimately be. The only successful

way to proceed, he asserted, was to shape the forms to the existing stone: "Sculptured work on a mountain must *belong* to the mountain as a natural part of it; otherwise it becomes a hideous mechanical application." Once the head of Washington was complete, he would determine how best to blend the next head with it.

He had chosen Mount Rushmore partly because the close-grained rock appeared to be carvable. It was immensely tough, but even so had a weathered surface which had to be cut away to expose smooth undamaged stone suitable for sculpture. For the first head, that of the first President, George Washington, Borglum cut away about 30 feet of rock. The most recessed head of the group, Theodore Roosevelt, required the removal of 120 feet of rock.

The problem of creating convincing faces for the four presidents was solved by a simple method invented by Borglum. First he prepared models, one-twelfth the size of the final sculpture, so that one inch on the models represented one foot on the mountain. At the centre of the head of each model he mounted a swivelled pointer, with a protractor plate to measure the exact angle, to right or left, to which it was pointing. From the pointer hung a plumb line, which could be moved in and out along the pointer and raised or lowered. Every point on the face of the model could then be defined by the angle of the pointer, the position of the suspension point, and the vertical fall of the plumb line. A similar pointer, but much bigger, with an arm 30 feet long, was then mounted at the centre of what was to be the head on the mountain. Measurements taken from any point on the model could be transferred to the same point on the rock, and marks made to show how much rock had to be removed. Men were trained to use the system, which proved simple and effective. It was the only measuring system needed to complete the entire sculpture.

The men Borglum hired to do the work were miners and quarrymen, familiar with pneumatic

Mount Rushmore
(5,725 feet) dominates the surrounding terrain and provided Borglum with smooth-grained granite that faced east, the best direction for the fall of light on the carvings. The final order of George Washington, Thomas Jefferson, Theodore Roosevelt and Abraham Lincoln was achieved pragmatically, beginning with Washington.

Builders of a Nation

drills and explosives, but hardly used to working hanging like spiders on the face of a mountain almost 6,000 feet high. To reach their place of work, they strapped themselves into devices rather like a child's swing, and walked backwards over the cliff as a winchman wound out the cable supporting them. To gain any purchase when drilling into the rock, they would first attach two bolts and a chain, which they put around their backs so that they were able to press against it. The rock was so hard that drill bits quickly became blunt, and a full-time blacksmith was employed to sharpen them.

To begin shaping each face, the drillers first created egg-shaped volumes of clean rock, with the surface 3 to 6 feet proud of the final profile. Then the pointers got to work, transferring to the rock the instructions for shaping it. By honeycomb drilling and chiselling, the rough outlines of the face were created, and finishing touches added at the instructions of Borglum, who had a genius for recognizing what was needed to make the faces come to life. Inspired touches were the way in which he suggested Lincoln's beard by vertical lines in the rock, and Roosevelt's spectacles by the bridge over the nose and just a hint of the outline of the frame around the eyes.

Work on the monument went on throughout the 1930s, with frequent pauses when money ran out or the weather proved too unpleasant. By the time Borglum died, on 6 March 1941, Mount Rushmore was all but finished. The final touches were carried out by his son Lincoln Borglum, who had started work as a pointer on the project at the age of 15. The final cost was just under £1 million. In only one place—on Jefferson's upper lip, where an area of uncarvable feldspar was found—did the sculpture need patching. A small piece of granite 2 feet long by 10 inches wide was pinned into place and cemented with molten sulphur. Today some 2 million people a year come to the Black Hills to see the Mount Rushmore National Memorial, justifying all of Doane Robinson's original hopes.

To celebrate the monument's fiftieth birthday in 1991, a renewal scheme costing $40 million is being undertaken. This will include the first structural analysis of the sculptures, to ensure that cracks in the rock are not growing. If repairs are needed they will be carried out, and the survey will provide other knowledge needed to manage what the Mount Rushmore superintendent Dan Wenk calls "this great resource".

The greatest alteration made by Borglum was to relocate the head of Jefferson after work had begun. As seen from the road, it was to have been to Washington's left (below) but the rock was unsound and Borglum disliked the perspective. However, a crack in the rock through Jefferson's nose in the new location forced him to tilt the head back.

Jefferson's rough-hewn face shows the technique of "honeycombing" in which a series of close-set holes was drilled and the honeycombed rock between removed with a chisel. The granite proved so hard that Borglum had to abandon his intention not to use dynamite, since the use of pneumatic hammers would have taken decades. After honeycombing, lighter pneumatic hammers, known as "bumpers", were used to chatter against the rock, smoothing the outlines and creating details.

Jefferson's complete head, *with Washington beyond, before work began on Roosevelt or Lincoln. Jefferson is the only one of the four to be portrayed as he looked before becoming president. The number of men working on the monument varied according to the weather and the availability of funds; it was sometimes as few as one, sometimes as many as 70 but an average was about 30. Work began at 7.30am after the climb up 760 steps to the mountain top. Supplies were brought in by the cableway anchored beside Jefferson's head.*

Borglum's models *were used to transfer his design on to the mountain using his "pointing machine" that related to a swivelled pointer in the centre of the head of each model (above).*

The pointer *(right) had a protractor plate to measure angles, while a plumb line which could be moved along the pointer, raised or lowered, enabled measurements to be transferred.*

Monument to Victory

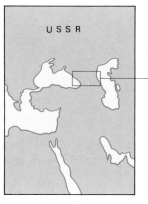

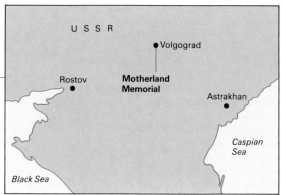

Fact file

The world's largest full-figure statue

Sculptor: Yevgeni Vuchetich

Built: 1959–67

Material: Reinforced concrete

Height: 270 feet

In the winter of 1942–43, one of the decisive battles of World War II was fought on the banks of the Volga at Stalingrad. German forces had broken through in August 1942 along a 5-mile front north of Stalingrad, and the Soviet 62nd Army faced encirclement and destruction by the German 6th Army, commanded by General Friedrich Paulus, and the 4th Panzer Army.

What followed was one of the most remarkable defences ever mounted, a battle in which every person available was mobilized to defeat the enemy. Both attackers and defenders endured appalling hardship, and terrible casualties: between February 1943, when the battle was won, and April 147,200 German dead and 47,700 Russians were buried. Of the 48,190 houses standing at the start of the battle, 41,685 were destroyed by bombing, fire, or artillery bombardment.

Some of the toughest fighting of all took place on a small hill to the north of the central part of the city. Mamayev Hill, named after the Tartar Khan Mamai who once made his camp on it, was listed on the combat maps as "Height 102"—its elevation in metres. The fight for possession of this crucial vantage point lasted for more than four months at the end of 1942, and it was here, on 26 January 1943, that units of the Soviet 21st Army advancing from the west joined up with the 62nd Army which had borne the brunt of the defence of the city. The German troops were cut in two, and defeated.

So intense was the fighting that the hill itself was altered in shape, and despite the severity of the winter it remained black, for the snow was melted by the heat of gunfire. When spring came, no grass grew. Every handful of earth from Mamayev Hill contained seven or eight pieces of shrapnel.

This hill still dominates the city, whose name was changed to Volgograd after Stalin's death

and posthumous disgrace. Now it is the site of the grandest war memorial in the Soviet Union, with its centrepiece a huge sculpture in reinforced concrete representing Mother Russia, somewhat sparsely clad, calling upon her sons to rise in her defence. The Motherland statue, created by the sculptor Yevgeni Vuchetich, is the largest full-figure statue in the world. From the base of her pedestal to the tip of the sword she holds aloft, Motherland stands 270 feet tall. The sword alone, made from stainless steel, is more than 90 feet long and weighs 14 tons.

The statue is certainly impressive, a landmark from every point in the city, but it is only one element in a majestic war memorial that is meant to be experienced as an unfolding drama. The whole concept was created by Vuchetich after he emerged in 1959 as the winner of a competition to design a fitting memorial for the dead of Stalingrad. It took 8 years to complete and was not finally opened until 15 October 1967.

Vuchetich, who died in 1974, was a prolific creator of sculptures glorifying the Soviet people and their triumph in the Great Patriotic War. He carried out more than 40 busts of generals, officers and soldiers, and at least ten Soviet cities have monuments by him. He was also responsible for a magnificent memorial to the Soviet Army in Treptow Park in East Berlin. Born in 1908 in Dniepropetrovsk, Vuchetich was educated at the Rostov School of Arts and at the Academy of Arts in Leningrad. He fought in the war, and suffered shell-shock. Afterwards he worked in the Grekov Studio of Painters of Battle Pieces.

The memorial he designed for Volgograd is didactic and uncompromising, a Politburo speech in stone. It begins at the foot of the hill, on Lenin Avenue. A stone mural depicts a procession making its way up the hill, the faces of men and women etched with grief but full of Socialist determination as they carry flowers, wreaths and banners to honour the memory of the dead. At the head of the procession are a man, his hand stretching out in the direction of the hill, and a girl carrying a modest bunch of flowers. They point the way to a flight of stairs that leads to a gently rising avenue of poplar trees. As soon as you set foot on the pathway, the huge Motherland figure is visible on the summit of the hill. She is standing against the wind, her scarf blown to one side. She seems to be shouting something, and pointing toward the Volga. The message needs no interpretation; she is calling on

Monument to Victory

her children to defend their country.

But before the huge figure is reached, you pass a smaller one, of a soldier emerging from a pool of water. Stripped to the waist and holding a grenade in his right hand and a sub-machine-gun in his left, this idealized member of the Red Army is no stripling, for he stands a full 40 feet high. Vuchetich called this sculpture "Fight to the Death" and said that it represented the whole Soviet people preparing to deliver a devastating blow to the enemy. "His figure, hewn from the beetling rock, becomes, as it were, a mighty bastion against Fascism" said Vuchetich, whose prose had something in common with his sculptures.

Behind this symbolic chunk of Socialist Realism, Vuchetich had the interesting idea of building two huge walls, converging in perspective and intended to convey the idea of a massive ruin. Like the rest of the memorial, they are carried out on a gigantic scale, more than 160 feet long and nearly 60 feet high. The walls, blackened by fire, are covered with inscriptions and scenes of the fighting. "Forward, only forward!" reads a typical one. At the end of the right-hand wall a real incident in the battle is depicted. A young member of the Komsomol (the Communist Youth Organization), Mikhail Panikakha, with no grenades left, is said to have destroyed a German tank by flinging himself on it with a flaming Molotov cocktail in his hand. Both he and the tank were consumed in the ensuing inferno.

The next stage in the ascent is Heroes' Square, a large open space with surrounding walls containing more depictions of heroism. Yet further and you reach the Hall of Military Glory, grey concrete walls outside but an enormous glittering hall within. The walls shine with gilt and copper inlay, and in the centre of the hall a huge hand faced with marble chips holds up a torch which carries the eternal flame. The torch is inscribed with the words "Glory, Glory, Glory", and the flame is surrounded by a guard of honour from the Volgograd garrison.

Perhaps uncertain that the many reliefs, inscriptions and sculptures would, unaided, make their effect on the Soviet public, Vuchetich and his colleagues also use sound to establish the appropriate mood. At the Ruined Walls can be heard the music of Bach, and the sound of wartime songs, the crash of gunfire, the shouts of the soldiers and the crackling voice of a radio announcer. In the Hall of Military Glory, it is

The Memorial Complex took over 8 years to build, entailing the excavation of over 1.3 million cubic yards of earth and the laying of over 26,000 cubic yards of concrete. To the left of Motherland is "The Grief of a Mother", which helps to balance the mass of the Hall of Military Glory on the opposite side of the square. The mother bending over the body of her lifeless son echoes the image of Pieta, expressing the human relationship between Christ and the Virgin.

Schumann's "Traumen", solemn and sad.

Around the Hall of Military Glory (also called the Pantheon) is yet another square, this time called the Square of Sorrow. Here there is another statue, "The Grief of a Mother", showing a woman bending over the body of her dead son. Finally there is another short climb to the base of the pedestal of Motherland herself. Along the concrete pathway that meanders its way up the grassy slopes are graves of Heroes of the Soviet Union who fell in the battle. "To Senior Lieutenant of the Guards Khazov Vladimir Petrovich, Hero of the Soviet Union, Eternal Glory!" "To Master Sergeant Smirnov Pavel Mikhailovich, Hero of the Soviet Union, Eternal Glory!"

As the path wanders to and fro, the statue is seen from different vantage points. Finally you are right beneath her feet and as you look up you notice, in the words of one Soviet writer, that the

The statue of **Motherland** (left) is not fixed to its pedestal, its own great weight providing the only support. The scarf blowing away behind the neck alone is said to weigh 250 tons. The statue is imposing from all angles.

colossal statue with her widely outstretched and highly raised arms has encompassed half the sky. Glinka's hymn "Glory" plays gently in the background. On winter evenings the statue is illuminated by searchlights.

The whole ensemble is typically Soviet, representing one of the principal justifications claimed by the Soviet Communist Party for ruling the country—victory in the Great Patriotic War. For sincere Communists, and for veterans of the war, visiting Motherland is an emotional experience. For younger more cynical Russians, it tends to be seen as a grandiose production of the years when Leonid Brezhnev led the country into stagnation. Motherland's local nickname is "Brezhnev's Auntie". But nobody should complain: in an earlier era it would not have been a statue of Mother Russia brooding over Volgograd, but of Stalin himself. And that would have been a lot harder to swallow.

"Fight to the Death" (above) is on the central axis of the complex and is made of a solid block of waterproof reinforced concrete faced with granite slabs. Birches, the typical tree of Russian forests, surround the pool from which the statue arises.

The Hall of Military Glory (left) is decorated with 34 mosaic banners with black ribbon fringes that bear the names of 7,200 Soviet soldiers who fell at Stalingrad. The floor is inlaid with black, grey and red marble.

CHAPTER TWO

Architectural Achievements

Great buildings are designed to make a statement. Some glorify God, or symbolize the power of a ruler—temples, cathedrals and palaces have been built ever since man first laid one stone on another. Some are monuments to wealth, or instruments of war; others are shrines to culture or to sport. Several reflect a desire evident through the centuries to build ever higher, creating structures which have tested contemporary building technology to the limit—and sometimes beyond it. Many towers have collapsed when ambition has outstripped knowledge of the laws governing stresses; one of the most remarkable towers to have fallen was the Gothic spire at Fonthill, which is featured in the Gazetteer.

Although the purpose of these buildings may vary, without a function they could hardly have been created, for architects, unlike other artists, cannot work without a client ready and willing to pay the bill.

All the buildings described here have some claim to be unique. They are either the first, the biggest, the tallest, the most original or the most fantastic of their kind. Some can claim, like the Crystal Palace, the Eiffel Tower or the New Orleans Superdome, to have carried the art of building into new territory. Others have been chosen because they reflect the obsessions of a single man, like Antonio Gaudí or Felix Houphouet-Boigny, determined to leave a statement in stone or concrete behind them. There are mysteries like the great pyramid of Cholula, science fiction fantasies like Biosphere II, and near-follies such as the Sydney Opera House in which a beautiful idea demanded to be translated into reality, however difficult that turned out to be.

Every important culture has produced great buildings; sometimes it is the only thing they have left behind. Here are a selection of the most remarkable among them, the work of architects and builders over the past four thousand years. If architecture is frozen music—as a German philosopher once claimed—here are some of the loudest and the sweetest sounds man has ever contrived to make.

Architectural Achievements

Temple of Amun
Pyramid of Cholula
Pyramids: shrines of the ancients
Krak des Chevaliers
Vatican Palace
Forbidden City
Crystal Palace
Paxton's Influence
Sagrada Familia Cathedral
Gaudi's creative genius
Eiffel Tower
Eiffel's other works
Manhattan
Epcot Dome
Munich Olympic Stadium
Sydney Opera House
Superdome
CN Tower
The tallest towers
Sultan of Brunei's Palace
Basilica of Our Lady of Peace
Biosphere II

Shrine to the God of the Wind

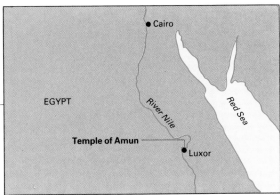

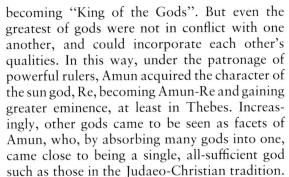

Fact file

The largest religious building ever constructed

Builders (major constructions):
Tuthmosis I—Ramesses II

Built: c1524–1212 BC

Material: Granite, sandstone and limestone

Area (Great Court):
10,668 square yards

On the banks of the Nile, at a place the Egyptians called the birthplace of all the world, lies the largest religious building ever constructed. The Temple of Amun at Karnak (once ancient Thebes) is more than a building, for its history stretches over 1,300 years. It is a record of Egyptian civilization, laid layer on layer in a vast and confusing muddle that impresses more by its bulk than its beauty. At its height, when Thebes ruled over Egypt, Amun's temple was served by 81,000 slaves, and was paid tribute in gold, silver, copper and precious stones from 65 other cities and towns. The many buildings on the site have one thing only in common: they were built to glorify the great god Amun, and to ensure their builders long life and great power.

The ancient Egyptians had many gods, and built shrines to propitiate them. Some gods had a purely local significance, but others were elevated to the status of "great gods"—like Re, the sun god, recognized as the source of life beyond his origins in Heliopolis, and Amun, the god of wind and of fertility, who was originally worshipped only in Thebes. With his wife Mut and son Khonsu, Amun formed a royal trinity,

becoming "King of the Gods". But even the greatest of gods were not in conflict with one another, and could incorporate each other's qualities. In this way, under the patronage of powerful rulers, Amun acquired the character of the sun god, Re, becoming Amun-Re and gaining greater eminence, at least in Thebes. Increasingly, other gods came to be seen as facets of Amun, who, by absorbing many gods into one, came close to being a single, all-sufficient god such as those in the Judaeo-Christian tradition.

The creation of the Temple of Amun coincided with the rise and fall of Thebes. Today all that is left is the temple itself, because the great city of Thebes was built, like all Egyptian domestic buildings, of mud bricks and has not survived. Even the houses where the Pharaohs lived were of brick, their furniture designed to last only a lifetime. The temple, however, was different. It was meant to last into eternity and was made of granite, sandstone and limestone, quarried and shaped with the most primitive of tools and techniques.

The granite came from quarries at Aswan, limestone from Tura near Cairo, and sandstone from many places along the Nile Valley. The softer stones appear to have been quarried with an implement like a pick, but no such surviving tool has ever been found. Stone slabs for building appear to have been dressed with a saw, probably made of copper and using an abrasive mineral such as quartz to increase its cutting power. Holes could be drilled with hollow circular drills, also made of copper; the cylindrical drill cores produced by the use of such an instrument have been found, although neither saw nor drill has survived.

The building methods used by the Egyptians were fairly primitive. The Temple of Amun, for example, has virtually no foundations. So long as the pillars could be laid on the underlying rock, that was considered sufficient. At Karnak, flooding carried away the flimsy foundations of the huge Hypostyle Hall, causing 11 columns to fall in 1899. This provided a chance to examine the foundations, which turned out to be little more than a trench packed with sand to provide a level surface, and a yard or so of small stones loosely laid on top.

The largest and most splendid building at Karnak is the Hypostyle Hall (from the Greek word meaning "below pillars") and consists of a forest of columns—originally there were 134 of them. This was the greatest building produced in

The Sacred Lake (left), bordering the south-east part of the temple complex, symbolizes Nun, the eternal ocean in which the priests of Amun purify themselves. The 134 columns of the Hypostyle Hall (right) are arranged in 16 rows, and those in the central double row are 69 feet high. Every surface is decorated with reliefs and inscriptions.

Shrine to the God of the Wind

ancient times, and covered an area 339 feet long by 169 feet broad. Down the centre runs a double row of columns 33 feet in circumference and 69 feet high.

To either side are seven further rows of columns, each 48 feet high. Originally the whole area, big enough to accommodate the Notre-Dame in Paris, was roofed with stone blocks, rising higher in the centre and with windows along the clerestory of the nave providing light for the interior.

Creating this huge building was an astonishing achievement, given the simplicity of the tools available. The pulley was not known to the Egyptians, and the blocks making up the pillars and the roof were pulled into place up ramps made of mud bricks. Scaffolding was used, but only on a small scale for the decoration and finishing of the stone. The men who built the temple worked in gangs with fixed shifts. A diary of the work done by each group was kept, together with records of the weight of copper tools issued to each man, and notes of excuses given for absence. The workmen were paid with food, wood, oil and clothing, and sometimes received a bonus of wine, salt or meat.

The Hypostyle Hall was planned and started by Ramesses I, who ruled for just two years before being succeeded by his son Seti I in 1318 BC. The hall was completed by Ramesses II, who succeeded his father in 1290 BC and ruled for 67 years. Ramesses II was a great builder, creating more temples and monuments than all the other Pharaohs, including the temples at Abu Simbel.

The decoration on the outside of the hall includes the depiction of Ramesses II's war against the Hittites, and includes the actual text of the final peace treaty, the first non-aggression treaty ever negotiated. It also includes a prayer to Amun for help when Ramesses II had been abandoned by most of his army and was faced by the might of the Hittites. "I call to thee, my father Amun. I am in the midst of strangers whom I know not. All the nations have banded together against me. I am alone and no one is with me . . . But I call and see that Amun is better for me than millions of footsoldiers and hundreds of thousands of charioteers." The northern wall depicts the battles of Seti I in Lebanon, southern Palestine and Syria.

The hall as we see it today is the result of major reconstruction, mostly by French archeologists. When it was rediscovered by Napoleon and his army at the end of the eighteenth century,

The use of calyx capitals to columns (left) provides a support 12 feet in diameter, enough to hold 100 men.

The decoration in the Hypostyle Hall has provided archeologists with valuable information. The Egyptians used little vaulting and no arches, their temples consisting of many pillars roofed over with flat blocks of stone. The limited span of limestone and sandstone necessitated numerous columns.

Ram-headed sphinxes (right) line the western approach to the Temple of Amun, from the River Nile. With sun discs on their heads and a statue of the Pharaoh under their chins, they symbolize the sun god's strength (the lion) and his docility (the ram).

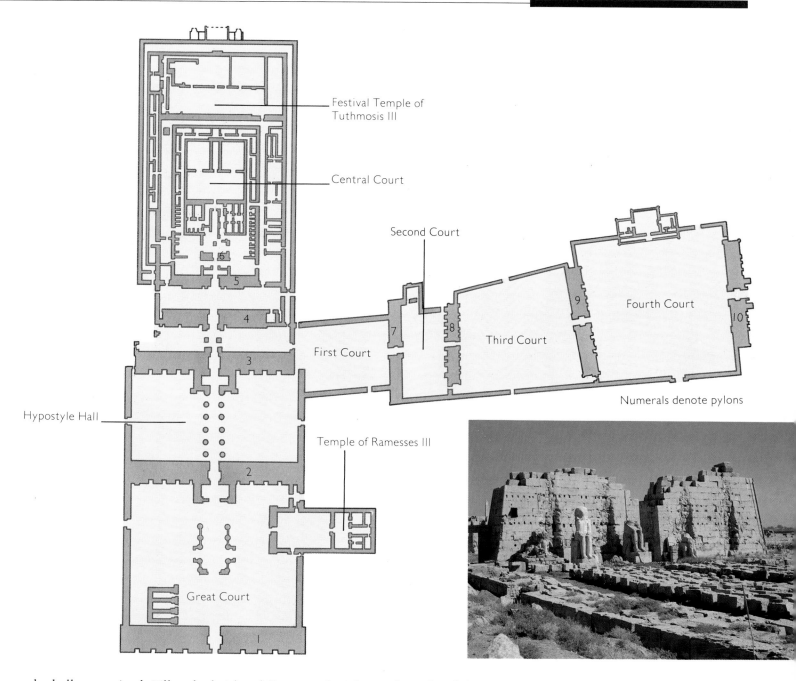

Festival Temple of Tuthmosis III

Central Court

Second Court

Fourth Court

Third Court

First Court

9

10

Numerals denote pylons

7

8

6

5

4

3

Hypostyle Hall

Temple of Ramesses III

2

Great Court

1

the hall was ruined. Pillars had either fallen or were leaning, the sand had encroached and almost buried it. Long years of reconstruction followed, to produce the complete building—less roof—that we can see today.

The Hypostyle Hall is just one of 20 temples, shrines and ceremonial halls at Karnak. The last structure to be built there, the giant pylon, or gates, was erected by the last native rulers of ancient Egypt, the Ptolemaic Pharaohs. This huge gateway is 49 feet thick, 143 feet high, and 370 feet wide. One wall, incomplete, still shows the rough finish of the stonework before it was finally dressed, and still in place are the remains of the brick ramps up which the blocks were hauled to make the wall.

Through the gates lies an open court built by

the Libyan Pharaohs of the XXII Dynasty (945–715 BC) and on the southern wall of this court is one of the finest examples of an Egyptian temple, the temple of Ramesses III, consisting of a forecourt, pillared hall, and sanctuary.

The whole area of the site is sufficient to accommodate 10 European cathedrals. It is an imperial statement on behalf of a god whose temple, at its peak in the time of Ramesses III, controlled at least 7 percent of the population of Egypt and 9 percent of the land, 81,000 slaves, 421,000 head of cattle, 433 gardens and orchards, 46 building yards and 83 ships. Discoveries are still being made by archeologists here; as recently as 1979–80, a complete shrine came to light, which is one of the most important discoveries in recent years.

The 8th pylon was built by Queen Hatshepsut. Pylons were introduced at Karnak by Amenhotep III (1417–1379 BC) and guard each side of the entrance to a temple. The walls slope inward, and the façades were usually decorated with scenes of conquest by whichever ruler built the pylon. Grooves were incorporated for ornamental flag poles.

Legendary Mexican Tomb

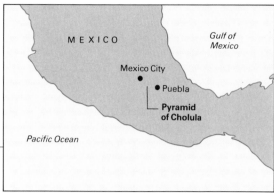

Fact file

The world's largest pyramid

Built: 2nd–8th centuries

Material: Adobe

Height: 200 feet

Base: 1,400 feet square

Near the quiet, sleepy town of Cholula in Mexico a pretty colonial church, its dome tiled in green and gold, sits on top of an odd-looking hill that rises from the plain. The church was built by the Spanish conquerors of Mexico, and is known as Nuestra Señora de los Remedios. It is one of many Christian churches in the town, but its builders may have been unaware that it crowns a much more remarkable religious structure. The hill on which it stands is not a natural feature at all, but the world's largest pyramid and the biggest ancient structure in the New World. It is, in fact, not one pyramid but at least four, each built on top of a previous construction.

The great pyramid, 1,400 feet along each side and some 200 feet high, was already in ruins and covered in dense green brush when the Spaniards first arrived. Building probably began on the site in the first or second centuries AD, and the successive increases in the size of the pyramid continued until the end of the eighth century, although modifications were made up until the twelfth century. Many thousands of people must have been involved in building such a huge structure, under the command of a priestly caste which exercised complete power. The earlier, buried pyramids at Cholula are contemporary with two other great pyramids at Teotihuacán, a bigger city 100 miles to the north and at its height the capital of a considerable empire.

The great pyramid at Cholula is built of adobe—unbaked brick—faced with small stones and then coated with either plaster or clay. Inside the pyramid there is a network of tunnels, in which many of the walls bear paintings, and a staircase of cut stone that leads through the inside of the pyramid to the flat summit. Outside the pyramid is a plaza more than an acre in area, which originally gave access to the staircase which led up the pyramid face.

Around the plaza are buildings, some with murals similar in style to those found at Teotihuacán, but at least one in a different style, namely a mural more than 150 feet long showing a ceremonial drinking scene that might have taken place at the time of the harvest. The figures, depicted life-size in a free-flowing style, are all men, with the exception of two wrinkled old women. The scene is one of abandonment, with the drinkers, mostly nude, showing distended stomachs which suggest that imbibing had been under way for some time. The mural is believed to have been painted between the second and third centuries.

Like the people of Teotihuacán, the god worshipped at Cholula was Quetzalcoatl, a creature with the feathers of a quetzal bird and the body of a snake. The feathers of this bird, which lives in a small area on the border between Mexico and Guatemala, were greatly prized in ancient Mexico for their rarity and beauty, so that the word "quetzal" eventually came to mean anything precious.

But who were the people who built the world's greatest pyramid? Nobody knows. They predated the Toltecs, who took over the region after their decline, and the Aztecs. But little is known about their language, customs, or the extent of their political control in the centuries during which the pyramid was being built. The huge size of the structure, and the organization that went into creating it, suggest that the society was controlled by an élite who commanded loyalty over a considerable area.

The better preserved ruins at Teotihuacán provide some clues about Cholula, for it is clear that the two places were linked. The city there was laid out on a grid plan, covering 8 square miles. Its main thoroughfare, the Avenue of the Dead, started at a huge pyramid, the Pyramid of the Moon, and passed in front of an even larger one, the Pyramid of the Sun. Along the length of

Legendary Mexican Tomb

this avenue were further pyramid-shaped platforms with flat tops, with a temple on each.

The huge Pyramid of the Sun, which rises to a height of 216 feet from a base about 750 feet square, is reckoned to have taken 30 years to build, using a work force of 3,000 men. About a million cubic yards of material was used to create it. The pyramid at Cholula, though not quite as tall, has a base almost four times as large, and is reckoned to have involved the shifting of 4.3 million cubic yards. Thus we may perhaps guess that it would have taken as many as 10,000 workmen a total of 40 years to build.

In practice, however, the building took place in stages, with the smaller pyramids providing a base for the later ones, so it is probable that the structure was created over a period of hundreds of years, with pauses as each successive pyramid was completed. For comparison, the Great Pyramid of Cheops, one of the original seven wonders of the world, was originally 481 feet high (now 449 feet, since the loss of the point at the top) and its base was 756 feet in each direction. It contains a total of 3.36 million cubic yards of material.

The people who built the pyramids at Cholula and Teotihuacán had only simple stone-age tools but were able to use them to create not only the monumental architecture of the pyramids, but also pottery and sculpture. On the east side of the great plaza at Cholula, archeologists found a huge stone slab, weighing 10 tons, carved along its vertical edge with a motif of serpents entwined with one another. On the west side of the pyramid, a stylized serpent's head was found, carved in a rectilinear style.

The makers of these artefacts do not appear to have been warriors. Neither Teotihuacán nor Cholula have any fortifications, which may explain how the priestly civilizations that created both disappeared so rapidly when nomadic warrior tribes arrived from the north. Teotihuacán at its peak was a city of at least 125,000 people, perhaps even as many as 200,000, which makes it larger than Athens at the height of its power. Yet it disappeared abruptly and completely in about AD 750 (some say earlier). Cholula was never as large, and may have survived a little longer, but it too was eventually overrun and its culture obliterated.

By the time Cortés arrived in Mexico, the city of Cholula had passed through the hands of at least three waves of conquerors. The best known of these were the Toltecs, who are said to have

taken control of the city in 1292, to be displaced in 1359 by the kingdom of Huexotzingo. Though neither of these peoples shared the religious convictions of those who built the pyramid, they continued to regard it as one of the wonders of the nation. Cortés himself reported that Cholula, a city whose exterior was "as fine as any in Spain", had 20,000 houses, and 400 pyramids.

At the beginning of the nineteenth century, the first tentative attempts to understand the ancient civilizations of Mexico were made by the German explorer and scholar Alexander von Humboldt. He was the first in modern times to measure the size of the pyramid, which he described as "a mountain of unbaked bricks",

The material used in much of the construction of the pyramid is adobe, which was faced with small stones and given a thick coating of plaster or clay. This render was then painted. The Spanish are said to have built 364 churches in Cholula.

Several stone monuments (left) have been found by excavation, though they were broken and have had to be repaired and re-erected where they once stood. This monument on the east side of the plaza is about 12 feet high with a missing top. Around the perimeter is an interlocking scroll design.

Tunnelling through the pyramid has helped to determine that there were at least 4 major superimpositions as the structure was steadily enlarged (above). The earliest measured 373 feet by 353 feet by 59 feet high, and the last phase increased it to 1,400 feet square and a height of about 200 feet, covering approximately 46 acres.

and he was struck by its similarity to the pyramids of ancient Egypt and to the Ziggurat of Belus in Babylon. He speculated about a link between the builders of these monuments.

Curiously, there is also a link between the pre-Conquest legends about the pyramid and those of the Biblical flood and the Tower of Babel. According to Humboldt, the pyramid was built after a great flood had devastated the land. Seven giants had saved themselves from the waters, and one of them built the pyramid in order to reach heaven. But the gods, angered by this plan, hurled fire at the pyramid in order to destroy it. Cortés was said to have been shown a meteorite, bearing a resemblance to the shape of a toad,

which had fallen on top of the pyramid.

A structure as large and as mysterious as the pyramid of Cholula is bound to attract myths, which can be dispelled only slowly by scientific efforts to understand the culture that created it. Excavations began at Cholula in 1931, and have continued since, with more than 4 miles of passages dug through the structure to expose its secrets. It is these excavations that have revealed the successive layers of pyramid-building, and cleared the platforms and squares of centuries of earth and dense greenery. But precise details about the people who built Cholula, how they did it, and why they disappeared so rapidly and so completely, remain to be discovered.

Excavations on the perimeter of the pyramid base have revealed a number of large plazas and courtyards surrounded by platforms. Protected by 30–35 feet of earth, they were in good condition, but the buildings had disintegrated. Only a small part of the site has been excavated.

Pyramids: Shrines of the Ancients

The earliest pyramids were constructed by the ancient Egyptians, and the first was the tomb built for the Third Dynasty king, Djoser (*c*2668–2649 BC). Its shape was almost an accident of its location; it required height to establish a dominant image so a square was enlarged, both at the base and with additional steps of brick. For the next millennium every king of note was buried beneath a pyramid. The largest was the pyramid of Cheops which held the record as the world's tallest structure for longer than any other—from *c*2580 BC until AD 1307 when it was overtaken by Lincoln Cathedral, England. The pyramids of central America were built much later, and were temples rather than tombs.

The Temple of the Giant Jaguar, Tikal, Guatemala
Situated in the largest of the Mayan cities, the pyramid is thought to date from about AD 800. The temple takes its name from the motif on a carved lintel. One side has been partially restored to reveal the succession of 9 terraces, surmounted by a 3-roomed temple. Within the temple a tomb was found; it contained a vaulted chamber with the remains of a skeleton adorned with 180 items of jade and buried with pearls, alabaster, pottery and shells. There are 7 other substantial pyramids in Tikal, which once covered 46 square miles.

The ziggurat at Ur
The ziggurat or temple tower is the principal ruin in the ancient city of Ur, in modern Iraq, which was inhabited c3500 BC. Ziggurats, like pyramids in central America, are thought to have been surmounted by temples, or to have encased burial chambers, as in Egypt.

The Pyramid of the Moon, Teotihuacán, Mexico
The pyramid covers 400 square feet at its base and provided the focus of the city, begun about AD 30. The other great pyramid in the city was consecrated to the Sun; Sun and Moon were represented by huge stone idols covered in gold.

The Pyramids at Giza, Egypt

Situated south west of Cairo is the world's largest pyramid, today measuring 481 feet in height—decay has reduced it from an original 449 feet. In its full form it contained $10\frac{1}{2}$ million cubic feet of material, and it is estimated that the pyramid contains 2.3 million blocks of limestone, each weighing $2\frac{1}{2}$ tons. The largest of 3 pyramids at Giza, it was built by Khufu (known to the Greeks as Cheops) who reigned in ancient Egypt from c2589–2566 BC. It was during the reign of his father, Sneferu, that the smooth-sided pyramid replaced the stepped kind.

The Soothsayer Pyramid, Uxmal, Mexico

The Mayan city of Uxmal is in northern Yucatan. The pyramid's total height is about 114 feet, made up of 4 distinct sections in an unusual elliptical form, surmounted by the base of a temple which is reached by 2 staircases. Restoration work has revealed that the pyramid has at least 5 building periods, reflecting the Mayan custom of superimposing new work on old. The name is derived from a legend about a dwarf whose soothsaying skill helped him to become king, whereupon he built himself this imposing pyramid.

Stronghold of the Crusaders

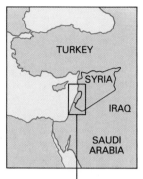

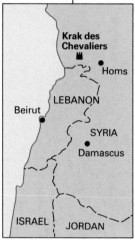

Fact file

The principal stronghold of the Crusaders which never fell to force of arms

Built: Successively strengthened between 11th and 13th centuries

Maximum wall thickness: 80 feet

Maximum garrison: 2,000

Headquarters of the Hospitallers: 1142–1271

Considered by many to be the finest medieval castle surviving today, Krak des Chevaliers is a reminder of the Crusades—religious passions that for two centuries drove thousands of men to war in a foreign land. The greatest fortress built by the Crusaders in the Holy Land, it was held by the Knights of the Hospital of St John for 130 years between 1142 and 1271. Standing on a spur dominating a fertile plain in what is now Syria, Krak was as nearly impregnable as any fortress ever built, falling finally as a result of a trick. T.E. Lawrence called it "perhaps the best preserved and most wholly admirable castle in the world".

Like other Crusader castles, Krak was built to defend the conquests made by Christian armies that had travelled to Palestine at the end of the eleventh century to liberate the holy places from Moslem occupation. The Crusades were inspired by Pope Urban II at the Council of Clermont in 1095, when he promised that Crusaders would be absolved from sin, would gain great wealth if they lived, or go straight to heaven if they died. Christ would be their leader in a Holy War. The effect was electrifying. "Never, perhaps, did the single speech of a man work such extraordinary and lasting results" one historian has written. The warlike sentiments of an increasingly confident and adventurous European nobility were sanctioned by the religious purposes of the Crusades.

After the First Crusade had succeeded in recapturing Jerusalem in 1099, many of the Crusaders left for home, their vows fulfilled. But some stayed, creating Crusader states along a narrow strip around the eastern shores of the Mediterranean. To protect these states from Moslem attack they built castles; the greatest of them was Krak des Chevaliers. Its name is a mixture of Arabic and French: Kerak, the Arabic for fortress, was corrupted into Krak, and the Chevaliers were the Knights of St John, who took over an earlier castle on the same site in 1140 and greatly improved it.

Krak was one of a network of Crusader castles standing on mountain peaks from the borders of Syria in the north to the deserts south of the Dead Sea. They were usually no more than a day's ride apart, and they were able to signal to one another at night by fires set on the battlements. They had their own water supplies, either in reservoirs cut from the rock or from natural springs, and they could withstand a siege for months, even years. They provided a system of defence which enabled the Franks and their successors to hold off a far greater number of Moslems for two centuries.

Krak was situated in the Crusader state known as the County of Tripoli, first established by Raymond of St-Gilles, Count of Toulouse. He died in 1105 and his successors first took possession of a small castle on the site, called the Castle of the Kurds, in 1110. They quickly improved it, but by 1142 the ruling Count of Tripoli, perhaps finding the responsibility for such an important castle too much for him, gave it to a religious military order, the Knights of St John, or Hospitallers. This order had run a hospital for pilgrims in Jerusalem and had earned the gratitude of Crusaders. Rewarded by the warriors whose wounds they had treated, the Knights of St John became a powerful and wealthy organization. It was during their occupation that Krak became the foremost castle in the Holy Land. Most of the work was done after an earthquake in 1202, which destroyed some of the existing fortifications.

Krak's plan is concentric, with two circles of walls interspersed with a series of towers. The masonry is massively solid, and the design epitomized the concept of defence in depth which reached its finest expression in this building. The succession of walls is designed to prevent surprise attacks, and to keep the siege instruments of attacking forces far enough away to prevent them from reaching the heart of the castle. The walls are made of masonry blocks 15 inches high by as much as a yard long, with a core of rubble and mortar, as was the usual medieval practice.

Below the three towers of the keep is a massive sloping wall, the talus, which drops more than 80 feet into a moat which also served as a reservoir. The angle of the wall, which Arabs called "the mountain", is puzzling, because its slope appears to make it easier for besieging forces to climb it. When T.E. Lawrence visited Krak in 1909 he was able, barefoot, to climb more than half way up this wall. He argued that its purpose could not be to prevent attackers undermining the walls, since the castle stands on rock, or as protection against battering rams, since its thickness—80 feet—would have been excessive. The reason must have been to prevent attacking troops getting so close to the wall that they were protected from the defenders' fire.

The same purpose is served by the placement of machicoulis around many of the castle's walls.

These are small boxes projecting from the walls close to the top. The object is to provide a secure point overlooking besieging troops from which fire could be directed at them, or stones or hot materials dropped on their heads. The machicoulis at Krak are small, meaning barely 16 inches across, and would have been big enough only for a single soldier.

To prevent attacking forces from attempting to rush the gateway and burst through, the entrance passage twists and turns through three "elbows"—abrupt changes of direction which make a blind charge impossible. The entrance is also defended by a drawbridge, a moat, four gates, a machicoulis, and by at least one portcullis.

In the days before the use of gunpowder, Krak was impregnable. During the occupation of the Knights of St John, a garrison of about 2,000 people was quartered here. A windmill stood on the north wall, grinding corn. Meetings and banquets were held in the hall, built in the thirteenth century, while the Latin Mass was chanted daily in the Chapel. The Wardens of the

Krak stands on a steep hillside at a height of 2,300 feet with a commanding view. Most of the towers are round rather than square, to minimize damage from siege catapults. The arched entrance and square tower beyond are of post-Crusader construction.

Stronghold of the Crusaders

The three tallest towers comprise the keep, built at the only point exposed to direct assault—the south. Previously keeps had been sited at the strongest point of defence, which was recognized as a tactical error. Beneath them is the sloping talus, thought to have been designed to keep attackers away from the wall and so make them an easier target.

An aerial reconstruction from the north east (right). In the foreground is the main gate; the chapel tower is that immediately beyond the main gate, with two lancet windows. The windmill was for grinding corn. Towers could not be roofed over to protect defenders, due to lack of wood or slate.

castle occupied rooms in the south-west tower, the same rooms in which T.E. Lawrence found the governor of the province living, with his harem, when he visited Krak in 1909.

There were assaults, but they failed. In 1163 the Emir Nur ed-Din besieged Krak, but made the mistake of taking a siesta one day outside the walls. The Knights poured out, surprised him, and put his army to rout. A generation later the great military leader Saladin marched his army up to the walls, took a good look, and retreated without even attempting a siege.

As time passed, however, the power of the Crusaders in the Holy Land began to fade. One after another the fortresses fell: Jerusalem in 1244, and Antioch in 1268. Krak found itself increasingly surrounded by hostile forces, growing every day more confident. In 1268, the Grand Master of the Knights of St John wrote to Europe for help, declaring that the bastions of Krak and Markab between them had only 300 men left to defend them against the Saracens. No help came, and in 1271 Sultan Beibars surrounded the castle with his army, and managed to penetrate the first defensive walls. But the talus, and the huge towers, defeated him. Within their walls the Knights could have held out for months.

Finally, Beibars devised a ruse. A letter, cleverly faked so as to appear genuine, was delivered to the defenders of the castle. It purported to come from the Count of Tripoli, and it instructed the garrison to surrender. The Knights emerged from their stronghold, and Krak fell. They were given a safe passage to the coast and rode away, leaving their castle behind them. All that is left, as one writer put it, are "the shadows of the kestrels cruising above, and the sun-scorched stones".

South strongwork

Inner Castle

Outer Castle

Outer moat

Box-machilcoulis

Loopholes

Warden's tower

Wall-walk

Refectory

Chapel

The vaulting of the Cloister *reflects the quality of Krak's stonework. The chamber of the Grand Master of the Hospitallers in the Warden's Tower has particularly fine work— delicate pilasters, Gothic vaulting and a decorative frieze.*

The outer wall *was overlooked by the concentrically placed inner wall, built on higher ground to provide support to the first line of defence. Both dominated the surrounding plain.*

Escarpments

Main gate
Main gate to inner ward

51

The Sacred City

The Vatican, one of the smallest states in the world, contains the second largest and most magnificent church. It also boasts the world's most famous ceiling—in the Sistine Chapel—and the world's largest collection of antique art, in the Vatican Museum. It has, in addition, a huge and famous library. Nowhere else are so many treasures of the Renaissance collected within so small a space.

It was here that St Peter, Christ's apostle and the first of the popes, was martyred, probably in the year AD 67. He was buried by his fellow Christians in a simple tomb on the sloping side of the Vatican Hill. Above this tomb was later built a great basilica by Constantine the Great which, despite the depradations of Goths, Huns, Vandals and Saracens, stood for more than a thousand years.

By the time of Pope Nicholas V (1447–55) the old building was tottering, its walls bulging 6 feet out of true and apparently ready to fall at any moment. Nicholas decided to replace it with a new building, but little was actually done until the papacy of Julius II (1503–13). Julius resolved to build a new St Peter's that would "embody the greatness of the present and the future . . . and surpass all other churches in the universe". He chose as the architect Donato Bramante.

Bramante devised a building in the shape of a Greek cross, with four arms of equal length, crowned by a magnificent central dome. The foundation stones were in place by 1507, and by 1510, 2,500 labourers under Bramante's direction had completed the four colossal piers that determine the size of the central crossing point. Bramante died in 1513, and in 1514 Pope Leo X appointed the young Raphael as chief architect. Raphael had already completed the decoration of the Vatican's State Apartments—the Stanze—with a series of paintings. He did not, however, contribute very much to the design of St Peter's before he died in 1520 at the age of 37.

Raphael was an agreeable man who obviously tried to make life pleasant for others, sometimes with alarming results. He allowed the masons working on St Peter's to leave holes in the foundations, to provide storage space for their lunches, tools and firewood. A few years later the hollow sections began to crumble and had to be replaced with solid masonry to make them strong enough to hold up the weight above.

Progress was slow after Raphael's death and stopped altogether during 1527, when Rome was sacked by invading Spanish forces. By the time work resumed in the 1530s, the piers finished so long before by Bramante were sprouting a lush growth of grass and weeds. The plans were changed by Antonio de Sangallo in the 1540s, but after a series of rows and the death of Sangallo and his successor in 1546, the elderly Michelangelo, then 71, was summoned by Pope Paul III to take charge.

Reluctantly Michelangelo assumed full responsibility, then worked on the building without pay for the rest of his life. He demanded *carte blanche*, and got it. He could make any changes he wished, even demolishing parts of the basilica which were already finished, and could draw on

Fact file

The greatest concentration of Renaissance art in the world

Architect, St Peter's Basilica: Bramante, Michelangelo

Built: 1507–1612

Materials: Stone and brick

Length: 694 feet

Area: 54,402 square feet

The Vatican City State (left) covers an area of 108 acres and contains 30 streets and squares, 50 palaces, 2 churches beside St Peter's, a radio station, a railway station and a printing works. In the right foreground is Castel Sant' Angelo, built by Hadrian in AD 130.

The Papal Altar seen from the principal entrance to St Peter's. Above the altar is Bernini's baldacchino, 95 feet high and supported by 4 gilded spiral columns, and above that Michelangelo's dome with an internal height of 452 feet and a diameter of 157 feet.

The Sacred City

money without having to keep formal records. The St Peter's of today is largely the achievement of Michelangelo, who shrugged off a torrent of criticism and several attempts to discredit him by jealous rivals. By the time he died in 1564 he had devoted 17 years to the building, under five popes. And by then the huge drum which carries the dome was complete.

It took another 26 years to complete the dome, after endless further delays. It was not until 1590 that the final stone was in place and Pope Sixtus V was able to offer a solemn Mass of Thanksgiving in the basilica. The dome as completed is not quite what Michelangelo intended, being taller and more pointed than his design.

Models show that the structure was also changed. Michelangelo's design consisted of three brick shells, one inside the other, but the finished dome has only two. Its structure consists of 16 stone ribs, with the gaps between them filled with bricks, laid in herring-bone patterns. Three rows of windows admit light into the

St Peter's Square (above) is an image known throughout Christendom for the Pope's weekly Angelus blessing and seasonal messages. Bernini's colonnade of 284 columns and 88 pillars in 4 rows carries an Ionic entablature and balustrade on which stand 140 statues of saints.

The dome of St Peter's is surmounted by a lantern, a copper ball 8 feet in diameter, and a cross. A complex of staircases runs through the dome and lantern into the ball itself.

The Pinacoteca (left) is the Vatican's picture gallery, comprising 15 rooms in the Lombardi Renaissance style and opened in 1932. Napoleon compelled Pius VI, who formed the collection of Old Masters, to surrender the best works to France in 1797, although 77 were recovered in 1815.

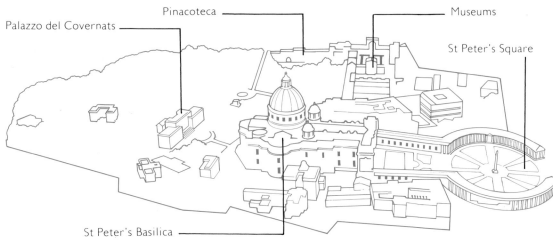

Palazzo del Covernats

Pinacoteca

Museums

St Peter's Square

St Peter's Basilica

The view from the dome of St Peter's (right) across Bernini's piazza and down Via della Conciliazione, a creation of Mussolini which entailed the demolition of 2 streets lined with old houses. At the end of the street is the Ponte Sant' Angelo, which also carries statues by Bernini—10 huge angels set up on the balustrade in 1668.

The Vatican Museums (above) contain the largest collection of art and antiquities in the world, with several thousand items of sculpture and 460 Old Masters. The main part of the collection dates from the time of Pope Clement XIV (1769–74) and is made up of Etruscan, Egyptian and Greek antiquities, many of which were found in Rome. Besides the 15 rooms of the Picture Gallery is the Museum of Modern Religious Art, housed in 55 rooms under the Sistine Chapel. The Vatican Library holds some 800,000 books, 80,000 manuscripts and over 100,000 engravings and woodcuts.

The Sacred City

A spiral staircase with barely perceptible risers links the entrance to the Vatican Museum, flanked by Statues of Raphael and Michelangelo, with the galleries above. From the main entrance of the museum to the Sistine Chapel—the highlight of any visit to the Vatican—is a walk of about ½ mile down marble corridors.

space between the two shells, and a narrow staircase runs upwards to the massive lantern on the top. Three chains were built into the dome during construction to prevent it spreading under the weight of the lantern, but 150 years later it became clear that this was not enough. Cracks began appearing as the ribs bulged outwards. Five more chains were inserted in 1743 and 1744, and a sixth in 1748. Since then there has been no further sign of movement.

In 1598, Clement VIII engaged Guiseppe Caesari to design the mosaics that decorate the inside of the dome, depictions of Christ, the Virgin Mary, apostles, saints and popes. Through the *oculus* at the very centre can be seen God bestowing blessings on mankind. Around the lower rim is an inscription in dark blue letters 5 feet high: *Tu es Petrus et super hanc petram aedificabo ecclesiam meam et tibi dabo claves regni caelorum* (Thou art Peter, and upon this rock I will build my church, and I will give unto thee the keys to the Kingdom of Heaven).

Even now the basilica was incomplete. The problem was that as designed by Michelangelo it did not include the whole area covered by Constantine's church, large parts of which had yet to be demolished. Should land hallowed by so many centuries of worship be allowed to fall outside the perimeter of the new building? Pope Paul V decided to extend the nave, turning

The splendour of the Vatican is reflected in the elaborate decoration of even a waiting room in the Secretariat of State. These walls were painted by Raphael, who was appointed Superintendent of Roman Antiquities by the Medici Pope Leo X (1513–21).

Bramante and Michelangelo's Greek cross into a Latin cross. Carlo Maderna designed the new nave, and an army of 1,000 men working night and day finished it by 1612.

One final detail remained, and it is the one that makes St Peter's instantly recognizable—the semi-circular colonnades that surround the huge Piazza di San Pietro, designed by Bernini and completed by about 1667. There are 284 columns, in the Doric/Tuscan order, and 88 pillars.

St Peter's was the work of many hands, over more than a century and a half. The ceiling of the Sistine Chapel, by contrast, was the work of one. Goethe said that nobody who had not seen the Sistine Chapel could have a complete conception of what a single man can accomplish. That man, of course, was Michelangelo, commissioned in March 1508 by Pope Julius II to paint the 12 apostles on the ceiling of the chapel. He agreed, reluctantly, and decided to do much more than that; to cover the whole ceiling, 10,000 square feet, with a huge fresco, a medium with which he was not even especially familiar. He called for assistants: seven applied, but after a short trial

all were sent on their way. Michelangelo locked the door and started the task alone.

He worked lying on his back on a scaffolding, paint dripping in his eyes and hair, and insistently nagged by Julius who kept demanding when he would finish. "When I can" Michelangelo replied. Working conditions were so uncomfortable that Michelangelo found he could not read a letter unless he held it above his head and tilted his head back. The job took him four years to finish, and then he signed it not with his own name, but with an inscription giving the honour of its completion to God—the alpha and omega, through whose assistance it had been begun and ended. The result was one of the great triumphs of the Renaissance, a joyous and glorious fresco that has been admired unstintingly ever since.

While Michelangelo was at work in the Sistine Chapel, Raphael was decorating the state rooms in the Vatican Palace. The four *Stanze*, as they are called, were the apartments used by Pope Julius II. The frescoes Raphael executed here are among his greatest works.

The Sistine Chapel takes its name from Sixtus IV (1471–84) who rebuilt it as the private chapel of the popes. Its chief glory is the barrel-vaulted ceiling frescoes by Michelangelo who, between 1508 and 1541, painted what has been described as a poem on the subject of creation, based on figures from the Old and New Testaments. Restoration of the frescoes began in 1980 and work on the main part was completed 10 years later amidst fierce controversy over its merits: opinion ranged from "an artistic Chernobyl" to high praise.

The Imperial Labyrinth

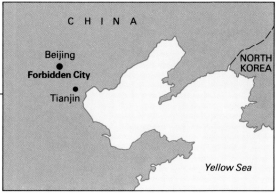

Fact file

For centuries the most mysterious and awe-inspiring palace in the world

Builder: Yung Lo

Built: 1406–20; mostly rebuilt

Materials: Wood and tiles

Number of rooms: 9,000

At the heart of Beijing (Peking) is the place "where earth and sky meet, where the four seasons merge, where wind and rain are gathered in, and where yin and yang are in harmony"—the Imperial Palace, or Forbidden City. In this huge complex of buildings the Ming emperors and their successors, the Manchus, ruled China for 500 years, attended by concubines and eunuchs and a few trembling bureaucrats who had to turn the orders of the "Sons of Heaven" into action. No ordinary citizen was allowed to step within its walls.

The Forbidden City that can be seen today lies on a site originally chosen by the Mongol rulers of the Yuan Dynasty (1279–1368) but was laid out by the third Ming emperor, Yung Lo, who ruled between 1403 and 1423. He came to power after a revolt against the grandson of the first Ming, Hung Wu, who has been described as "the harshest and most unreasonable tyrant in all of Chinese history". Hung Wu's violent temper and unreasonable cruelty so terrified his officials that if called to an audience with the emperor they would say their last goodbyes to their families.

When Hung Wu died he was succeeded briefly by his 16-year-old grandson, who was soon overthrown by his own uncle. Despite his name, which means Everlasting Happiness, Yung Lo was as tough, cruel and capricious as Hung Wu. He decided to shift the capital of China from Nanjing, closer to his own power base in northern China. In 1404 he began the reconstruction of Beijing, and the bulk of the Forbidden City was created between 1406 and 1420, using up to 100,000 craftsmen and as many as a million labourers. Its construction was one of the greatest building feats in history.

The plan to which Yung Lo built was said to have been given to him in a sealed envelope by a famous astrologer. It is based on geomantic principles, with each important building repre-senting a part of the body. It lies along a single straight line, the axis of the Universe, in which the emperor's role was "to stand at the centre of the Earth and stabilize the people within the four seas", according to the Confucian sage Mencius. The main axis runs north and south, with a series of courtyards and pavilions succeeding one another in rigid sequence. The whole area covers about 250 acres, and is surrounded by a moat and also by a wall 35 feet high, with four doors.

The city is divided into two sections, with the buildings of state (including six main palaces) in the first and the residential buildings behind. The whole includes 75 halls, palaces, temples, pavilions, libraries and studios, linked by courtyards, paths, gardens, gates and walls. Altogether, there are reputedly a total of 9,000 rooms.

The Forbidden City was built not in stone, but wood. As a result, its buildings deteriorated or were destroyed far more quickly by fire, rot and insect attack than if they had been of more permanent materials. Of the buildings that stand today in the Forbidden City few are very old by the standards of Europe. Many were destroyed when the city was sacked and looted by the Manchu armies at the overthrow of the Ming Dynasty in 1644, and were rebuilt by the Qing emperor Qian Long (1736–96). Further additions were made by the Empress Dowager Cixi during the nineteenth century. Why the Chinese emperors did not choose to build more permanently is not clear, for their own mausoleums were built in stone. The most persuasive explanation is that the emperors were more concerned with the life eternal than life on Earth, and therefore devoted greater energy and resources to creating enduring structures that they would occupy after death.

Architecturally, two things are particularly striking about the Forbidden City: the exotic curves of the roofs, and the brilliant colour of the

buildings. Although their method of constructing roofs could have been adapted to build in planes rather than curves, it seems that the Chinese preferred curves for aesthetic reasons. They enjoyed the contrast between the straight lines of the pillars and the base of the buildings and the languorous curves of the roof.

Entering the Forbidden City through the Wumen, the Meridian Gate—originally reserved for the use of the emperor—there is a huge courtyard. From the heights of the gate the emperor would review his armies, survey prisoners to determine who should live and who should die, and announce the new year's calendar to the court. His power was so absolute that

it was for him to designate the days and months of the year. When inspecting his troops, he would be flanked by elephants provided by his Burmese subjects.

Beyond this courtyard and through the smaller Taihamen (Gate of Supreme Harmony) lies a second, larger court in which the major imperial audiences would be held. The whole court of perhaps 100,000 could be accommodated here, and they would enter through the side gates—civilians to the east, soldiers to the west—before standing in silence before the emperor and prostrating themselves nine times.

Facing them as they gave obeisance was the first of three main ceremonial halls, set one

The courtyard between the Meridian Gate, the southerly entrance to the Imperial City, and the Gate of Supreme Harmony (on the left) is the first of several large open spaces between the principal halls. Between the parallel balustrade runs a stream crossed by 5 bridges, intended to symbolize the 5 virtues.

The Imperial Labyrinth

The view from Prospect Hill, looking south towards Meridian Gate and Tiananmen Square, shows the uniform sweep of the roofs and the scale of the city. In the foreground is the principal northern gate, the Gate of Divine Military Genius.

behind the other on a raised marble terrace called the Dragon Pavement. The Taihedian (Hall of Supreme Harmony), originally built in 1420 and restored in 1697, is the largest building in the Forbidden City, covering more than half an acre and standing 115 feet high. No building in the whole of Beijing was permitted to be higher. It was used for special state occasions such as the emperor's birthday. Here was his throne, with two elephants of peace at its feet and a screen behind symbolizing in dragon motifs both longevity and the unity of earth and heaven. Twenty columns support the roof, the six in the centre being decorated with the imperial dragon.

In the second, smaller hall, the Zhonghedian (Hall of Perfect Harmony), the emperor prepared himself and put on the imperial regalia for these ceremonies, while the third, the Baohedian (Hall of Protecting Harmony), was used for Palace Examinations, the system by which candidates for positions in the administration were chosen. The principle was to pick candidates by merit: the origin of all modern meritocracies. In practice, there was much corruption, and the examinations became increasingly formalized, demanding only the learning by heart of the tenets of Confucius. In this hall, the Emperor also received rulers bringing tribute. The anterooms to the hall have now been converted into galleries to display imperial relics and the gifts given by foreign rulers—many still in their original wrappings, conveying the Chinese contempt for the tribute of barbarians.

Beyond the three great halls, in the Inner Court, are the buildings where the emperors lived. The first, the Qianqinggong (Palace of Heavenly Purity) was the residence of the last four Ming emperors. The last, the Kunninggong (Palace of Earthly Tranquillity) was where their empresses lived, and where the emperor and empress traditionally spent their wedding night. The actual wedding chamber, a small room painted entirely in red with decorative emblems symbolic of fertility, was last used in 1922 for the child wedding of Puyi, the last Manchu emperor. Between these two halls was the Jiaotaidan (Hall of Union, or Vigorous Fertility) which was used for birthday celebrations, and for storing the seals of previous emperors. On display here

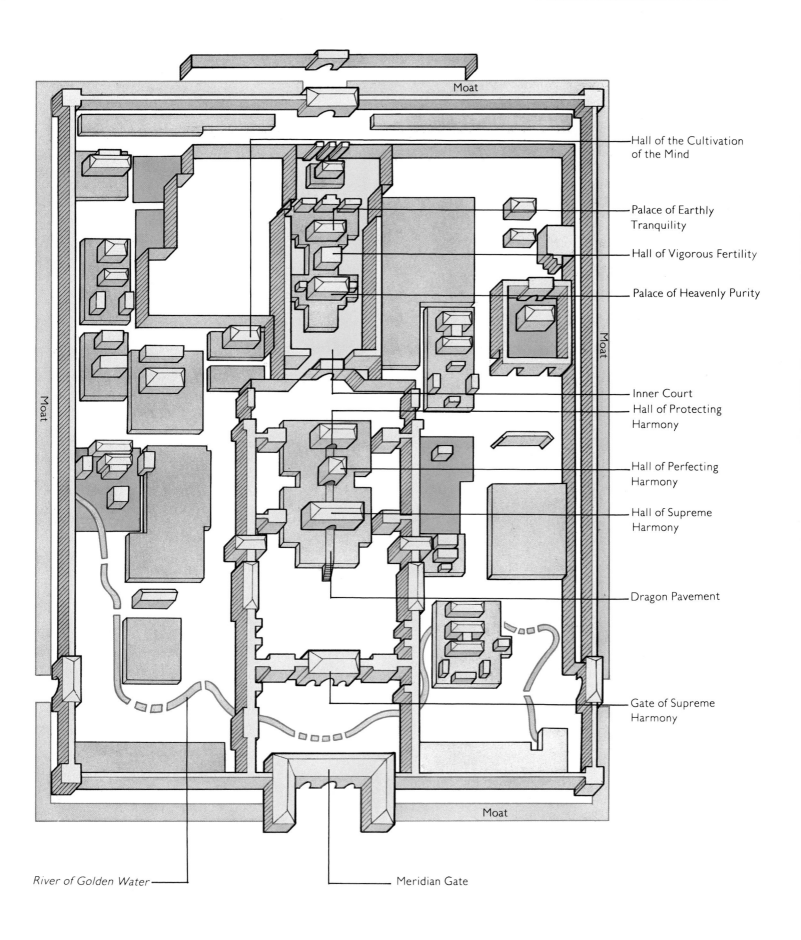

Moat

Hall of the Cultivation
of the Mind

Palace of Earthly
Tranquility

Hall of Vigorous Fertility

Palace of Heavenly Purity

Moat

Inner Court

Hall of Protecting
Harmony

Hall of Perfecting
Harmony

Hall of Supreme
Harmony

Dragon Pavement

Gate of Supreme
Harmony

Moat

Moat

River of Golden Water

Meridian Gate

The Imperial Labyrinth

The Dragon Pavement leads through the 3 main halls between the Gate of Supreme Harmony and the Gate of Heavenly Purity. Steps flank a bas-relief of dragons which was reserved exclusively for the imperial palanquin. The last imperial occupant of the palace was Puyi, subject of Bertolucci's film The Last Emperor.

today is one of China's inventions, the clepsydra (water clock), 2,500 years old.

The Hall of Heavenly Purity, which is surrounded by a complex of houses, medical consulting rooms, libraries and living quarters for palace servants, was the place where the emperors left the instructions concerning the succession. Each emperor would write the name of his chosen successor on two slips of paper, keeping one and concealing the other behind a plaque on the wall inscribed with the words "Upright and Bright". On the emperor's death his advisers would recover the two slips and compare them. If the same name was found on both, the chosen successor would be named.

These six halls form the main north–south axis of the Forbidden City. Their purpose was principally ceremonial, and the emperors actually spent most of their time in another building to the west, the Yangxindiang, or Hall of the Cultivation of the Mind. Within the entire complex they lived their whole lives, seldom setting foot outside among their people. Few ruling families have been more remote, autocratic, or self-indulgent. Their meals were gargantuan, their sexual appetites satisfied by hundreds of concubines, and the daily routines of the palace carried out by eunuchs, the only male attendants allowed to live within the Forbidden City.

The role played by the eunuchs during the Ming dynasty became increasingly dominant. They were employed because it was thought they would be loyal and reliable, having no families of their own, and no prospect of illicit relations with the palace women. Most of them were recruited from the ranks of criminals who had been castrated as a form of punishment; pathetically, since the Chinese belief was that no person who was not whole could aspire to heavenly happiness, they would carry their scrotums around with them, or at least ensure that they were buried with them when they died. Hung Wu had tried to limit their number to 100, but by the end of the Ming era in 1644, there were 70,000 eunuchs in the Forbidden City, and another 30,000 on administrative duties outside.

As the Ming emperors declined in vigour, the power of the eunuchs increased. By the 1620s, the power of the government fell first into the hands of a concubine, and then into those of a 52-year-old eunuch, Wei Chung-hsien. Wei became so influential with the emperor, a 15-year-old boy whose principal interest was carpentry, that he became the virtual ruler. Temples were erected in his honour, while opponents were executed in numbers "beyond calculations", according to the official history of the Ming. Wei lost his power only when the emperor suddenly died, and he was forced to commit suicide to avoid arrest. The Ming dynasty soon fell to the Manchus, who burned part of the Forbidden City and melted down the silver.

In the nineteenth century, the Forbidden City did fall under the power of a concubine, the autocratic Empress Dowager Cixi. Her power derived from the fact that of all the emperor Hsien Feng's concubines, she was the only one to provide him with a son and heir. He died when the child was only five, and Cixi assumed power, fighting off other courtiers. When her son died at 19 she struck again, insisting on the appointment of another young emperor so that her regency could continue. When this emperor took the throne and began to introduce reforms, Cixi struck for a third time, returning from semi-retirement to take power once more.

Narrow-minded, brutal and xenophobic, Cixi made common cause with the members of a group called the Society of the Righteous and Harmonious Fists, who blamed foreign imperialists for China's ills. When the society—known to Westerners as the Boxers—attacked missionaries, Cixi refused to accede to Western

The use of colour throughout the Forbidden City is determined by the different elements of the buildings: the raised podiums upon which they are built are white; the pillars and walls are dull red (far left); and the roofs are brilliant golden yellow (left). This colour was reserved exclusively for imperial use.

demands for their suppression. In June 1900 they attacked foreign-owned buildings and legations in Beijing, and swung Cixi behind them. In August, Western forces arrived to rescue the besieged diplomats, invading the Forbidden City and putting Cixi to flight. But disagreements among the Western forces enabled her to return, and she arrived once more in Beijing in January 1902. Finally she attempted to introduce the reforms that 30 years earlier might have preserved the dynasty as a constitutional monarchy, but it was too late. In 1908 she died, and in 1911 the revolution led by Sun Yat Sen triumphed.

Attempts to restore imperial rule during the 1920s failed, and in the 1930s many valuable objects were looted from the Forbidden City under the Japanese occupation. The retreating forces of Chiang Kai-shek took away still more in 1949, as they abandoned mainland China for exile in Taiwan before the Communist forces of Mao Zedong. On 1 October 1949 Mao stood on the terrace of the Gate of Heavenly Peace and proclaimed the birth of the People's Republic of China: the latest dynasty to rule China.

Today the Forbidden City provides the backdrop to the mass rallies held in Tiananmen Square, which were particularly associated with the Cultural Revolution and the cult of Mao whose portrait has been incongruously hung on the Gate of Heavenly Peace.

The intricate carpentry under the eaves (above left) was purely ornamental. On some buildings it became so elaborate that an extra colonnade had to be placed under the outer edge to support the weight. Ferocious bronze lions flank the Dragon Pavement (above right).

Paxton's Inspiration from Nature

Fact file

The world's first exhibition building constructed of glass and iron

Designer: Joseph Paxton

Built: 1850–51

Materials: Wrought and cast iron, glass

Length: 1,848 feet

Width: 408 feet

Few buildings have been designed so swiftly, or erected at such breakneck pace, as the Crystal Palace. It took less than a year from the moment Joseph Paxton conceived the building to the day in 1851 when it was opened by Queen Victoria to house the Great Exhibition. More than twice the size of St Paul's Cathedral, it covered 19 acres of Hyde Park, and in its central transept a full-grown elm tree 108 feet high was comfortably accommodated. The design called for 4,500 tons of cast and wrought iron, 6 million cubic feet of timber and 300,000 panes of glass, and was itself revolutionary, paving the way for modern steel-framed buildings. The actual construction of the building took just seven months.

Paxton formulated his brilliant idea at exactly the right moment. The Great Exhibition was intended to demonstrate Britain's pre-eminence in engineering, and its principal patron was Prince Albert. A distinguished Royal Commission had been established to plan and organize the Exhibition, and had delegated the decision on the building to a committee of engineers and architects which included Charles Barry, architect of the Houses of Parliament, Isambard Kingdom Brunel, and Robert Stephenson. After wading through 245 sets of plans submitted to them, they were at their wits' end. In desperation they produced a design of their own—largely the work of Brunel—and, equally desperately, the commissioners accepted it. It was awful, envisaging a huge, squat, brick warehouse with a vast iron dome on the top. It would have required at least 16 million bricks, and even if enough could have been found, it is doubtful if there would have been time to lay them all. The plans were greeted with horror by *The Times* and many shared its views.

Into this frenzied atmosphere stepped Joseph Paxton, the Duke of Devonshire's head gardener. Paxton, born in 1803, was a farmer's son, with little formal education. The duke had

spotted his talents and hired him at the age of 23 to run the gardens at Chatsworth. There Paxton had worked wonders, digging lakes, diverting streams and shifting hills to beautify the duke's estates. There he built a lily house, using the principle behind the leaf structure of the giant water lily *Victoria regia* for the framework. It had not been finished long when Paxton decided to apply the same methods to the design of a building for the Great Exhibition. Although it was at the eleventh hour, the committee was prepared to consider his design, providing they had it in their hands within two weeks. "I will go home and in nine days time I will bring you my plans all complete," Paxton told them.

He went straight to Hyde Park to look at the site, and was confirmed in his decision to build a greatly enlarged version of his lily house. Such a design had many advantages: it would be quick to erect and, having no mortar or plaster, would be dry and ready for occupation; it could be taken down just as easily, and put up somewhere else, answering the critics who said the Exhibition was going to destroy Hyde Park; and if no permanent site could be found, at least the materials would be worth a lot as scrap.

During a railway board meeting three days later, an inattentive Paxton made his first drawings; although mere jottings, they contained the essence of the design, a rectilinear building rising in tiers. The pillars were to be of iron, the walls of glass. In a week, the plans were complete. It took another week for the building contractor, Fox & Henderson, and the glass manufacturer to produce precise costings. They determined that Paxton's building, with its 205 miles of sash bars, 3,300 iron columns, 2,150 girders and 900,000 square feet of glass, could be built for £150,000, or for £79,800 if they could keep the materials after the building had been taken down. The committee had no choice but to accept—even the higher figure was lower than the estimate for their own design.

Once construction began, the genius of Paxton's design became apparent. The iron columns, hollow to take the rainwater flowing off the roof, could be erected with remarkable speed and girders laid across the top. Once the workmen got into the swing of it, they could raise three columns and two girders in 16 minutes, as Paxton himself reported. As the first storey progressed, other teams came along behind to build the second. Special machinery on site made the miles of "Paxton guttering"—

The revolutionary techniques employed in the construction of the Crystal Palace perfectly complemented the idea of the Great Exhibition—to illustrate Britain's industrial pre-eminence. It was opened by Queen Victoria and Prince Albert (below).

Paxton's Inspiration from Nature

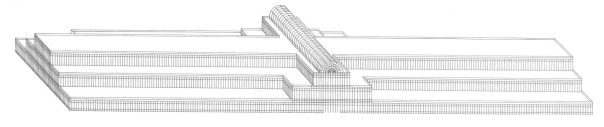

Paxton's design was derived from the lily house he built for the Duke of Devonshire, itself based on the principle of radiating ribs and cross-ribs which Paxton observed in a lily at Chatsworth.

wooden rafters hollowed out on top to act as gutters, with channels inserted on the underside to carry away water which condensed on the inside of the glass.

The arches of the transept, which transformed a huge glass box into an elegant building, were made of wood and were lifted into place from above. Once they were in position, glazing began. In one week, 80 men fixed 18,000 panes of glass. For their productivity the glaziers sought a rise from 4 shillings to 5 shillings a day and went on strike. Fox & Henderson reacted in typically Victorian fashion, dismissing the strike leaders and giving the rest a chance to go back to work—at the old rate. They did.

Everybody was astonished as the building rose so fast above Hyde Park. It was *Punch* that gave it its name—the Crystal Palace—and William Thackeray wrote a verse or two in celebration:

> As though 'twere by a wizard's rod
> A blazing arch of lucid glass
> Leaps like a fountain from the grass
> To meet the sun!

By now critics were fewer, although some proclaimed that a strong wind or a shower of hail would cause the building to collapse. *The Times* suggested that the official salute, due to be fired on the opening day when Queen Victoria arrived, would "shiver the roof of the Palace, and thousands of ladies will be cut into mincemeat". No such disaster occurred, and opening day, 1 May 1851, was an unparalleled triumph. "The sight as we came to the middle was magical," wrote Queen Victoria in her journal, "so vast, so glorious, so touching." By then the huge building had been filled with millions of objects, many of them attesting to the vigorous lack of taste of mid-Victorian Britain.

The exhibition proved a huge success, its profits being used to finance more permanent London buildings which grew into today's complex of museums between the Brompton Road and Hyde Park—the Victoria and Albert, the Science and the Natural History museums.

The speed of construction was a tribute to Paxton's genius: the use of shear-legs, pulleys and horses obviated the need for scaffolding; the girders bolted simply to the columns (below left).

More than six million visitors went through the turnstiles before the Exhibition eventually closed on 11 October.

Paxton was anxious that his masterpiece should survive and led a campaign to leave it where it stood in Hyde Park. But opposition was too strong, and Parliament rejected the proposal. But by then Paxton had raised £500,000 to buy the building and a new site for it, on 200 acres of wooded parkland on the summit of Sydenham Hill, on London's southern outskirts. Here it was rebuilt, even bigger and more splendid than before. The Sydenham Crystal Palace was half as large again as the one in Hyde Park, with a vaulted roof from end to end and a transept doubled in width. When built, it was filled with extraordinary objects: courts representing the different periods in the history of art, hundreds of sculptures—some of them colossal—trees,

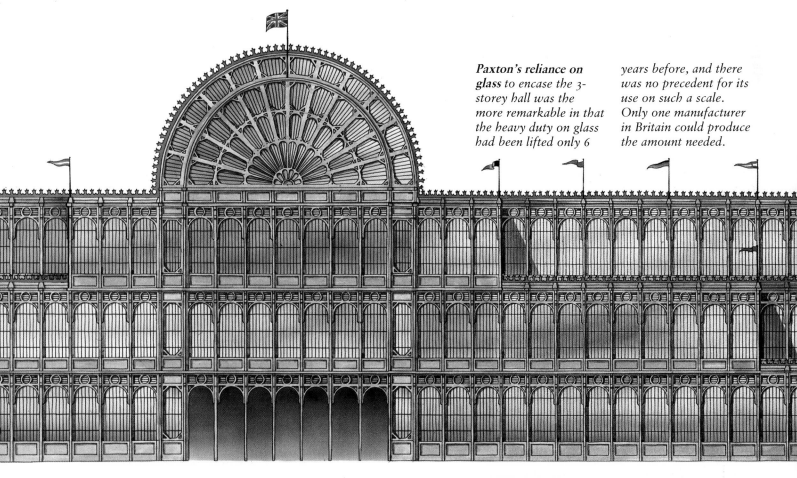

Paxton's reliance on glass to encase the 3-storey hall was the more remarkable in that the heavy duty on glass had been lifted only 6 years before, and there was no precedent for its use on such a scale. Only one manufacturer in Britain could produce the amount needed.

The transept running at right angles to the building was the brilliant idea of the exhibition committee. It was prompted by the strong opposition to felling trees on the site; a 108-foot high transept would enable a group of elms to be kept (left).

It took 900,000 square feet of glass to cover the Crystal Palace. The glass was $\frac{1}{16}$ inch thick and supplied by Chance Brothers of Birmingham. The glaziers worked from small trolleys with wheels that fitted into the gulleys on Paxton's gutters, propelling themselves along. Rainwater flowed down the gentle ridge-and- furrow roof into gutters and down the 8-inch hollow columns. The Times *warned that the* "concussion [of guns firing the opening day salute] will shiver the glass roof of the Palace, and thousands of ladies will be cut into mincemeat".

Paxton's Inspiration from Nature

The Sydenham Crystal Palace was not Paxton's preferred solution to the problem of what to do with the building when the Great Exhibition closed on 11 October 1851. He had hoped that it could remain in Hyde Park and be converted into a Winter Park and Garden with an abundance of trees and plants. Parliament rejected the idea, but Paxton had already raised £¼ million to buy a site and re-erect the building.

Besides creating at Sydenham Hill the intended botanical collection, Paxton had copied statuary, urns and vases from early civilizations, and built stupendous fountains that rivalled Versailles. The 2 towers at each end (above) were built by Isambard Kingdom Brunel to supply the necessary head of water. When completed 12,000 jets used 7 million gallons an hour.

art galleries, a hall of fame, a theatre, a concert hall with 4,000 seats and room in the centre for a Grand Orchestra of 4,000 musicians and a Great Organ with 4,500 pipes.

The Crystal Palace at Sydenham was not a museum, or a concert hall, or a huge park; it was all three at once, perhaps the first example of what are today called Theme Parks. A family could spend the whole day there, enjoying the setting and spectacle, and finishing in the evening with a huge firework display for which the place became famous. It was here, too, that a large audience first watched moving pictures. There were balloon ascents, high-wire acts, shows, exhibitions, conferences, pantomimes and spectacular events such as the staging of an invasion in which an entire village was destroyed in front of 25,000 spectators. The Crystal Palace provided for the first time a leisure centre where people from any background could enjoy their free time.

All this came to an end on 30 November 1936. A small fire broke out in a staff lavatory and, despite efforts to put it out, spread with alarming speed. The wood of the floorboards, the walls and the sashes burned fiercely, defeating the efforts of 89 fire engines and 381 firemen to put it out. It was the biggest and most spectacular event ever staged at the Crystal Palace, visible from all over London. People flocked to watch as the great building was destroyed. By morning it had gone. In the gloom of the 1930s there was never a serious chance it would be rebuilt.

Paxton's Influence

Joseph Paxton was born in 1803 at Milton Bryant, near Woburn in Bedfordshire, into a family in poor circumstances. Through hard work and intelligence, he came to the notice of the Duke of Devonshire; at 23 Paxton was put in charge of the Duke's gardens at Chatsworth.

The principles behind the construction of the lily-house at Chatsworth and the Crystal Palace were to have a profound effect. The train sheds at King's Cross, St Pancras and Paddington were derived from his work. Even more important was the establishment of the ideas of system building, and of the option to rely on an interior framework rather than an exterior wall for a building's strength.

The train shed for the Midland Railway's London terminus was designed by R.M. Ordish and W.H. Barlow, who had helped Paxton with the Crystal Palace. The lattice cast-iron ribs supporting the roof are tied together by girders under the platforms.

The Bond Centre, Hong Kong
Built by the Australian entrepreneur Alan Bond, this skyscraper office block is typical of the thousands of office buildings worldwide that employ glass as a curtain wall and are based on a system of prefabrication that reduces costs—and limits the freedom of the architect.

Willis Faber Dumas offices, Ipswich, England
Designed by Foster Associates and completed in 1975, this building exemplifies the reduction of the outer wall to a weathershield, having no structural function. Internal steel or concrete frames and replacement of window frames by silicone or neoprene joints enables the curtain wall to be made entirely of glass.

Gaudi's Gothic Masterpiece

Fact file

The world's most unorthodox cathedral

Architect: Antonio Gaudí y Cornet

Built: 1882–

Material: Stone, brick, steel and concrete

Height: 557 feet

Seating: over 13,000

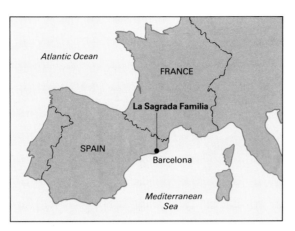

For more than a hundred years, a church has been under construction in Barcelona. It is huge, fantastic, and unfinishable, the dream of an architect whose imagination ran away with him. The cathedral of La Sagrada Familia—the Sacred Family—is a building unlike any other, where pillars lean and branch like trees, and huge pierced towers stand silent before an empty nave. It has been described as a work of genius, and as the product of a diseased imagination: few building sites have excited such strong and contrary emotions.

La Sagrada Familia began as a perfectly respectable neo-Gothic church, to be built in the "New City" area of Barcelona, and financed by the Spiritual Association of the Devotees of St Joseph. Its purpose was to exalt St Joseph and the Holy Family, symbols of family life and thus the basis of the social order. A site was bought, the local diocesan architect, Francisco de Paula del Villar, produced a design, and the foundation stone was laid in 1882.

Quite soon the architect fell out with the association and was replaced by a young man of only 31, Antonio Gaudí. What began for Gaudí as an architectural commission became a lifelong obsession, a devotion blending religious observance and art into a consuming passion. He never completed the building; it remains unfinished. But it is the greatest sight in Barcelona and one of the most extraordinary conceptions in the whole of Western architecture.

The style adopted by Gaudí for La Sagrada Familia is difficult to describe, for it has no exact counterpart elsewhere. It borrows from Gothic, but the curling, almost liquid shapes of the stonework owe a lot to Art Nouveau. It is as if the drawings of Aubrey Beardsley, or the silverwork of the English Arts and Crafts Movement, had been turned into stone. Gaudí's main

The site of the High Altar (left), placed beneath the central cupola. Gaudí's concept for the High Table envisaged Christ on the cross as the only ornament, with a vine winding around the foot of the cross. The 7 apsidal chapels would be dedicated to the joys and sufferings of St Joseph.

The Facade of the Nativity (right) on which work began in 1891. It took until 1930 to complete. The 4 bell towers are dedicated, from left to right, to the Holy Apostles Barnabas, Simon, Thaddeus and Matthias. Facing east the façade is lit by the rising sun.

Gaudí's Gothic Masterpiece

influences appear to have been John Ruskin and William Morris, and the French neo-Gothic architect Viollet-le-Duc. He worked on the church for so many years between the acceptance of the commission in 1883 and his death in 1926 that it also reflects his own changing views about architecture and religion.

Gaudí's first step was to make the church bigger. He would have liked to alter its position, too, but the foundations were already laid. For the first ten years or so he continued building the crypt, in more-or-less Gothic style, his main innovation being to introduce naturalistic ornamentation. But from the 1890s, his ideas blossomed. He abandoned Villar's plain ideas, substituting a profusion of decoration, with floral, human and animal motifs.

By 1895 he was designing the east façade, a controversial decision since the people of Barcelona were already becoming impatient, and the west façade, which faced the city, appeared a more urgent priority. Gaudí justified his decision by pointing out that the theme of the east façade was the birth of Christ, and thus it must be built before that of the west, whose theme was the Passion. Gaudí already saw the church not as a building to be finished as quickly as possible, but as a religious expression in its own right, a catechism in stone.

His plans grew ever more ambitious and complex. Around the church were to be 18 pointed towers with, at the centre, a great tower 557 feet high—as high as Cologne Cathedral and far higher than either St Paul's in London or St Peter's in Rome. Gaudí intended the towers to symbolize the 12 apostles, the 4 evangelists, the Virgin Mary and, the tallest of all, Christ himself. The three façades of the church Gaudí saw as representing the birth, death and resurrection of Christ.

This rich use of symbolism is also seen in the details of the design. Gaudí appears to have abhorred plain surfaces. What strikes the visitor is the sheer dynamism of the decoration, with animals, plants, figures, trees and sculpture occupying every square foot. Many of the sculptures would have been framed in colour if Gaudí had lived to see them completed. Around the whole building he planned a cloisterlike structure which would have shielded the inner sanctum from the noise of the street.

The four towers of the east façade, each 328 feet high, were the last parts of the church constructed under Gaudí's direction, and he

lived to see only the southernmost tower finished, which is dedicated to St Barnabas. Since his death and a long interruption from 1936—when the Spanish Civil War stopped work—to 1952, building has continued. But it is far from complete even today. However, an attempt is being made to complete the church in time for the Olympic Games at Barcelona in 1992.

The glory of Gaudí's conception can only be appreciated in a fragmentary way, as when the interior is illuminated at night, and the light floods out through the pierced stonework. Then La Sagrada Familia represents, as Gaudí hoped it would, the expression in masonry of Christ's words: "I am the light of the world."

La Sagrada Familia was in a part of Barcelona not built up until this century. The contrast between the hue of the stone on the Façade of the Nativity (above) and the newer towers behind, illustrates the effect of decades of urban pollution. The Coronation of the Virgin is above the central main door.

The interior of the Façade of the Nativity. The position of statues was determined by Gaudí pragmatically: full-size plaster models were put in place in the early morning light while he watched from a distance. They were then moved about according to his wishes. Gaudí was so concerned with day-to-day details that he often slept on site.

A detail of the Façade of the Nativity (below) illustrates the riot of decoration and symbolism that covers every surface of the exterior. Showing the section above the central doorway, which represents Charity, under the windows is the nativity scene. The musician-angel on the right replaces one destroyed in 1936 during the Spanish Civil War.

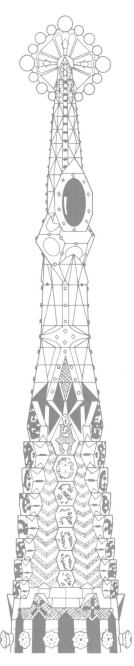

The pinnacles of the towers above the Façade of the Nativity are encrusted with mosaics and pearl shapes, and they represent the episcopal attributes of the cross, mitre, crook and ring. The words "Hosanna" and "Excelsis" are alternately rendered in vertical letters.

Gaudi's Creative Genius

Antonio Gaudí y Cornet, born on 25 June 1852 at Reus, Tarragona, was a fervent Catalan and a devout Catholic whose buildings are the artistic expression of the Catalan political revival. Despite a humble background, Gaudí's strong character and intelligence took him in 1873 to the Provincial School of Architecture in Barcelona.

Although staunchly Catalan—he refused to speak Castilian—he was conservative both in his private and his spiritual life. Gaudí neither married nor travelled, nor did he establish a school of architecture. His was a vision that died with him, at a moment when the geometric and functional International Style of modern architecture was in the ascendant.

Park Güell (right) was intended by Gaudí to be comparable to an English garden suburb, but only the park was built. He worked there until 1914 and it is now the city's public area, containing a church, arbours, sculpture and Greek Theatre. This is the Hall of 100 columns.

Casa Batlló in Barcelona (above) was a remodelling of an existing building. Between 1905 and 1907, Gaudí covered the plain building with a mosaic portraying sky, clouds and water, gave it a dragonlike roof of glazed scales, and inserted slender columns in the windows, prompting the building's nickname, "House of Bones".

The terrace of the open-air theatre in Park Güell (right) has a serpentine curlicue of benches made from shards of ceramic, glass and porcelain, and even the bottoms of plates and bottles from local potteries. Besides being cheap, the materials encouraged spontaneous composition. In his pragmatism, Gaudí was more like a medieval craftsman than an architect.

Gaudí's use of natural forms was influenced by the writings of the art critic John Ruskin. Rather than take elements from nature, Gaudí used them as whole as possible: besides this lizard in Park Güell (above), he incorporated in his work flowers, seeds, trees, snails, dogs, fish scales, bone and muscle.

Mosaics are used on the house in Park Güell to emphasize the building's form. The use of brightly coloured ceramic tiles was introduced to Iberia by the Arabs. Gaudí's love of ornament, colour and unusual forms testifies to a sense of humour as well as form. Gaudí is revered in Barcelona for the vigorous and sometimes playful buildings he bequeathed the city.

Stairway to the Sky

Fact file

The world's most distinctive tower, built to commemorate the centenary of the French Revolution

Designer: Gustave Eiffel

Built: 1887–89

Material: Wrought iron

Height: 990 feet

Eiffel's tower dominates the Paris skyline with an elegance that even its early critics finally admitted.

France's most instantly recognizable landmark, the Eiffel Tower, was denounced as an eyesore when first proposed. "A dishonour to Paris and a ridiculous dizzy tower like some gigantic and sombre factory chimney" declared a group that included the writers Alexandre Dumas and Guy de Maupassant and the composer Charles Gounod. Today it is impossible to imagine Paris without its "tragic lamppost", "inverted torch-holder" or even "Grand Suppositaire"—all descriptions applied to it at various times.

The tower was built for the 100th anniversary of the French Revolution, which was marked by a huge exhibition, the *Exposition Universelle*, in Paris. The organizers considered a number of schemes for a centrepiece for the exhibition, including the bizarre idea of a model guillotine 1,000 feet high. The best idea came from Gustave Eiffel, an engineer who was already well known as an expert in wrought iron. He had built bridges, domes and roofs using this reliable material, at that time cheaper than steel. The tower had been suggested to him by two juniors in his engineering company, Maurice Koechlin and Emile Nougier, who did the preliminary calculations. Eiffel took the idea to the organizers of the *Exposition* and convinced them to back the project.

The aim was to construct the world's tallest building, 984 feet high. At the time, the record was held by the Washington Monument in Washington, DC, a stone obelisk that was 554 feet high. The tallest ancient building was the Great Pyramid of Cheops, at 482 feet. Eiffel's target was to build a structure almost twice as high as anything that had gone before.

His design was a structure of wrought-iron ribs, held together by rivets and resting on a solid masonry foundation. Unlike a bridge, where a large number of the struts are identical, the tower required many different components, which were individually designed by a team of 50 engineers under Eiffel's direction. Each component had a maximum weight of 3 tons to ease construction.

The actual building of the tower began in January 1887. Steel caissons 50 feet long, 22 feet wide and 7 feet deep were filled with concrete and sunk into the subsoil to form the foundations. Upon these the wrought-iron structure began to rise at the end of June. The components were lifted by cranes, and so accurate was their manufacture that even when the tower rose 164 feet above ground the holes in the prefabricated parts matched exactly. This was important, because wrought iron cannot be welded; it must be fastened together with rivets. Once the first platform had been completed (by 1 April 1888), the cranes were hoisted on to it.

Work continued steadily throughout 1888, and by the end of March the following year the tower had reached its full height. One of the more remarkable statistics is that nobody was killed during the construction, although an Italian workman died while installing the elevators, after the tower had been inaugurated. The tower weighed 9,547 tons and was built of 18,000 components, held together with 2.5 million rivets. The workforce consisted of only 230 men, 100 of them to make the parts and 130 on site to put them together. The final height of the tower was 990 feet, or a fraction more in hot weather when it expands by a further 7 inches.

On 31 March a small party climbed the 1,792 steps to the top of the tower to hoist the French *tricolore*, a vast flag more than 23 feet long by 15 feet wide. Toasts were drunk in champagne, to cries of "Vive la France! Vive Paris! Vive la République!". The descent from the platform, *The Times* reported, "was found to be as trying as the ascent had been, and lasted 40 minutes". On the ground, tables were laid for a celebration attended by 200 workmen, the engineers who designed the tower, and the prime minister, M. Tirard, who admitted that he had not at first been enthusiastic about the tower, but was now prepared to make an *amende honorable* and concede that he had been wrong.

Stairway to the Sky

Now that the tower was up, many of its critics found it much more elegant than they had expected. It was lighter and more graceful than it had appeared in the drawings. Gounod retracted his criticisms and *Le Figaro* celebrated the tower's opening with a tribute in verse to its creator: "Gloire au Titan industriel / Qui fit cet escalier au ciel" (Glory to the industrial Titan who built this stairway to the sky).

Nor did the gloomy predictions of financial disaster come true. The tower cost 7,799,401 francs and 31 centimes to build—about 1 million francs more than Eiffel had predicted—but it attracted a huge number of visitors. In the last five months of 1889 alone it was visited by 1.9 million people, who paid 2 francs to get to the first platform, a further franc to get to the second, and another two to reach the top. By the end of the year, 75 percent of the total cost had been recovered. It went on to become a highly profitable enterprise, although the 1889 attendance was not exceeded until the advent of mass tourism in the 1960s. In 1988, a total of 4.5 million people visited it.

Originally designed to last for only 20 years, the tower is still going strong after 100. During the 1980s it underwent a major facelift, costing $28 million. One job was to remove excess weight which had gradually been added to the structure over the years. Some 1,000 tons, including a revolving staircase 590 feet high, were removed.

The Eiffel Tower has always been operated as a commercial enterprise. M. Citroën, the motor car manufacturer, temporarily owned the publicity rights and rigged up an imposing system of lights by which flames appeared to creep up from the base of the tower to the top. More usefully, the tower formed an excellent platform for radio and, later, television transmitters.

The tower is painted a muddy brown colour, which the French call *marron*. The paint's full name is *brun Tour Eiffel*, and 45 tons of it are applied to the structure every seven years. Inevitably, it has been the scene of many suicides; some 400 people have thrown themselves off it. The first parachute jump was made in 1984 when a British couple, Mike McCarthy and Amanda Tucker, slipped past security guards and jumped from the top, landing safely. An elephant once walked to the first platform, and two motorcyclists managed in 1983 to ride trail bikes up the 746 steps to the second platform, turn round and come down again without mishap.

Two elevators serving the first floor were to a French double-deck design to carry 50 people. During World War II a mysterious fault prevented Hitler from using them, compelling him to climb the tower on foot. When the city was freed in 1944, the turn of a screw released them.

Hydraulic jacks were incorporated into the base of the tower's 16 columns (4 for each pier). They enabled the piers to be adjusted so that they were perfectly horizontal for the band of girders situated at first-floor level.

The Stages of Construction

Prefabrication of sections was a revolutionary innovation, forced on Eiffel by the close deadline for completion. He decided that the 7 million holes in the girders would be drilled off site, leaving only riveting to be done in situ *using portable forges.* 5,300 drawings specified the location of holes.

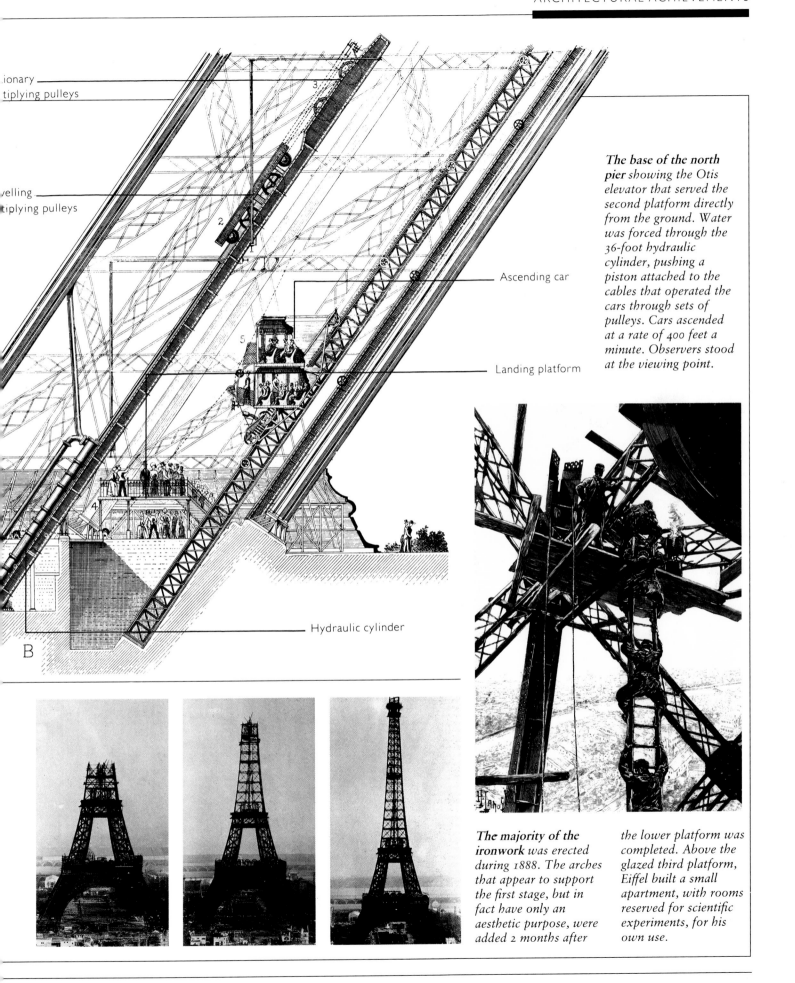

ionary
tiplying pulleys

velling
tiplying pulleys

Ascending car

Landing platform

Hydraulic cylinder

B

The base of the north pier *showing the Otis elevator that served the second platform directly from the ground. Water was forced through the 36-foot hydraulic cylinder, pushing a piston attached to the cables that operated the cars through sets of pulleys. Cars ascended at a rate of 400 feet a minute. Observers stood at the viewing point.*

The majority of the ironwork *was erected during 1888. The arches that appear to support the first stage, but in fact have only an aesthetic purpose, were added 2 months after* *the lower platform was completed. Above the glazed third platform, Eiffel built a small apartment, with rooms reserved for scientific experiments, for his own use.*

Eiffel's Other Works

Alexandre Gustave Eiffel was born in Dijon on 15 December 1832. After graduating in chemistry in Paris, he joined a company manufacturing railway equipment, which encouraged him to give up chemistry for civil engineering. At the early age of 25 he was put in charge of the construction of a bridge over the Garonne River at Bordeaux. He adopted a new method of pile driving, and his success in completing, on schedule, one of the largest iron structures of its day, helped to establish his name.

During a recession, Eiffel opted to become a freelance consulting engineer and soon set up a metalworks in Paris. His growing reputation led to contracts for bridges as far afield as Peru, Algeria and Cochin China as well as countless viaducts and bridges for railways in Europe. But his skill extended to all forms of engineering: a harbour for Chile; churches in Peru and the Philippines; gasworks, a steelworks and dam in France; and lock gates for Russia and the Panama Canal. The tower for the Paris Exposition was but the climax of a remarkable career.

Eiffel died at the age of 91 on 27 December 1923 at his mansion on Rue Rabelais, Paris.

Bon Marché Department Store, Paris, France
Eiffel's designs for ironwork were grounded in rigorously worked-out calculations which enabled him to build with the minimum of ironwork without sacrificing strength or rigidity. He even published a formula applicable to all wrought-iron structures which eliminated much of the guesswork from calculating stresses and strains. The lightness of his designs is evident in the Bon Marché department store in Paris (above and left) which Eiffel built with L.-C. Boileau in 1869–79.

The Nice Observatory, France

Situated in the Alpes-Maritimes, the Nice Observatory (left) was the largest such dome in the world when completed by Eiffel in 1885. Eiffel produced the ironwork of the dome, 74 feet in diameter, which rotated on a frictionless ring that enables the 110-ton dome to be moved by hand.

The Garabit Bridge, France

This viaduct in the Massif Central (below) was second only to the Eiffel Tower in Eiffel's pantheon of engineering achievements. When opened in 1884, it was the highest arched bridge in the world, at 400 feet above the Truyere River. The arch of 541 feet supported a railway deck 1,850 feet in length.

The Steel and Concrete Forest

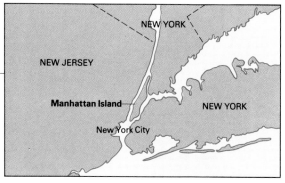

Fact file

The world's greatest concentration of skyscrapers

Length: $12\frac{1}{2}$ miles

Width (maximum): $2\frac{1}{2}$ miles

Length of electric cable under Manhattan: 17,000 miles

The island of Manhattan, $12\frac{1}{2}$ miles long by $2\frac{1}{2}$ miles wide, has the world's most sensational skyline. Here, in the heart of New York, huge buildings soaring almost out of sight are constantly being constructed, often to be demolished within years and replaced by ever-taller structures. Manhattan is never finished, for no sooner is a new building erected than the architects, engineers and builders move on to another site and start again.

Space is so limited that there is no alternative but to build upward, and as techniques have improved so the height of Manhattan's tallest buildings has increased. From the Flatiron Building of 1903 to the Empire State Building of 1931 and the World Trade Center of 1971, Manhattan has usually been able to boast the world's tallest habitable building. Even when that title has been wrested from it by an exceptional building elsewhere, Manhattan can claim to have the greatest concentration of skyscrapers in a single place.

Much ingenuity has gone into making it possible. Until Elisha Otis developed his "safety hoister", buildings were limited to the height people were prepared to climb on foot, and that was generally no more than about six storeys. Otis invented a safety device that locked his elevator in place even when the cable supporting it was cut, and showed it off at the New York World's Fair in 1854. Using cast-iron frames, it was already possible to build higher, and by 1875 the Western Union Building on Lower Broadway reached ten storeys. By the end of the century the Pulitzer Building on Park Row had exceeded it. Topped by a huge cupola, this building was a mixture of old and new; it had a core supported by wrought-iron columns, but its outer walls rested on masonry up to 9 feet thick.

Traditional buildings require thick walls to support their bulk; the taller they are, the thicker their walls must be at ground level. Given the limitations of space in Manhattan, that would have set an upper limit to the builders' ambitions had it not been for the evolution of the steel-framed building, in which all the internal and external loads are carried by the frame and transmitted by it to the foundations.

The first of these was William Jenney's ten-storey Home Insurance Building in Chicago, built in 1884. The first in New York was probably the Tower Building at 50 Broadway, designed in 1888 by Bradford Lee Gilbert, on a site only 21 feet wide. If he had followed traditional methods, all but 10 feet of Gilbert's building at ground level would have had to be solid masonry. Instead, he built what he described as "an iron bridge truss, stood on end"— a building 13 storeys high in which the iron framework went all the way up. To calm the fears of the owner, Gilbert himself promised to take the top two floors, so demonstrating his confidence that the building would stand. Like many of New York's pioneering skyscrapers, Gilbert's building has given way to something else, but his method opened the way to even taller structures.

One of the most striking is the Flatiron Building, situated on a narrow triangular site the shape of a flat-iron at the junction between Broadway and 5th Avenue at 23rd Street. Its 20

Midtown Manhattan looking west. The Chrysler Building is just to the left of the Empire State Building (the tallest skyscraper visible) and to the right is the white, angular profile of the Citicorp Center. The tower on the extreme right, with a pierced gable, is the A.T. & T. Building.

The Steel and Concrete Forest

The World Trade Center, built between 1966 and 1973, is an unusual development for the USA in that it was financed by two states, the Port Authority of New York and New Jersey. The intention was to bring together over 1,000 businesses and government agencies involved in international trade; over 60 countries are represented. Half the 16-acre site was reclaimed from the Hudson River, and the underground foundation work was 6 storeys deep. The tallest tower, of 110 storeys, is 1,350 feet high.

storeys were built on a steel frame clad with decorative stonework and served by six Otis hydraulic elevators. It was built by the George A. Fuller construction company, which boasted that it was the strongest building ever erected. Unlike many pioneering buildings it survives today, and is New York's oldest skyscraper.

But it was soon dwarfed by much bigger buildings of which the most striking was the 792-foot Woolworth Building, a Gothic tower on Broadway built in 1913 for F.W. Woolworth, founder of the chainstore that still bears his name. This has 60 floors from sub-basement to top, each floor with generous 12-foot ceilings—a use of space that could not be justified economically today. The inner structure of this building is steel, but its outer decoration is of terracotta, elaborately modelled into complex shapes and traceries. So profitable were the Woolworth stores that the building cost of $13.5 million was paid out of cash flow as construction proceeded.

The steel structure of the Woolworth Building, like all skyscrapers of the period, was riveted together. The rivets were inserted red-hot into prepared holes in the steel columns, and shrank as they cooled, tightening their grip and holding the steel pieces firmly together.

This was how New York's most famous skyscraper, the Empire State Building, was put together in 1930–31. When you stand on the observation tower 1,250 feet above 5th Avenue, it is disconcerting to think of the riveters who stood here on narrow steel beams, tossing red-hot rivets around with nothing but air beneath them. The building's graceful design, with higher levels set back by increasing amounts from the ground line, was the result of New York's building code of the time, which specified that a tower could not simply go straight up. The architect, William Lamb, produced 15 different designs before selecting the one that was finally built.

Work began at the height of the Depression and went on at a hectic pace; there were days when it rose by more than a storey. The building contains 60,000 tons of steel beams, produced in

The Empire State Building was the world's tallest building when built, in 1930–31, its 102 storeys reaching 1,250 feet. Designed by Shreve Lamb & Harman, its construction was so well planned and executed that it was built within 18 months and some occupants moved in 4 months ahead of schedule. The reason for the gradually recessed walls of the tower lies with New York's building code, which prohibited a straight rise from the street for more than 125 feet. A further reduction in plan was required at the 30th floor.

The Woolworth Building held the record for the world's tallest office building for almost 20 years after its completion, in 1913, at 761 feet. Designed by Cass Gilbert it remains one of the most highly regarded skyscrapers, and was given protected landmark status by New York City in 1983. So successful were F.W. Woolworth's 5c and 10c stores that the \$13½ million bill for construction was paid out of income. The building's ceiling heights—over 12 feet— are much more generous than today's developers would allow. The tower is covered in terracotta and all details reflect its Gothic intentions.

Pittsburgh and delivered on a relentless schedule which had many beams in place only three days after they had been made. The entire weight of the building is 365,000 tons.

The planning of the construction has gone into legend. Every day a progress chart and a printed timetable were issued. They listed every truck that would be arriving, what it would be carrying, who was responsible for it, and where it should go. Space in Manhattan is so tight that buildings can seldom afford the luxury of a builder's yard for storing materials during construction. Each steel piece was numbered to make sure it went into the correct place, and on each floor as the building rose a small gauge railway was built to carry materials to the right spot. Unloaded at ground level on to small carts, materials were lifted by derricks to the upper floor, placed on the track, and then wheeled to the precise spot they were needed.

A trip up the Empire State Building is one of the unmissable treats of a visit to New York, and two million people every year ride its express elevators which reach the 80th floor in less than a minute. Further elevators continue to the glass-enclosed upper observatory at the 102nd floor level. Above that rises a TV mast which itself is as tall as a 22-storey building.

Initially, the public doubted the strength of the structure, but all uncertainties on that score were removed in July 1945 when a US Air Force bomber approaching Newark airport in fog and rain smashed into the building at the 78th and 79th floors. The crew of 3 and 11 people inside were killed in the impact, but the building itself stood firm. The riveters had done their work well.

Today's techniques are rather different. Steel structures are held together by bolts, or by welding, rather than rivets. The men who dangle high above the streets putting them together are called ironworkers, or steel erectors. They are armed with spud wrenches, a tapered point on one end to jam in the holes to force a beam into position, a spanner on the other to tighten a nut on to a bolt to hold the two pieces together.

Other skyscrapers use no steel at all, but are made of concrete, poured on site into wooden moulds known as forms. The vertical pillars that separate the floors and hold them up are cast around a sprouting mass of steel reinforcing bars, while the floors themselves are made by pumping concrete on to a temporary wooden floor covered with a mesh of reinforcing bars.

The Steel and Concrete Forest

Between 4 and 8 inches of concrete are poured, a shallow edge around the outside of the building preventing its escape, and levelled.

Within a day or so the concrete is strong enough to walk on, and the wooden floor is removed and raised to the next level to create another storey. Temporary wooden beams may be inserted to hold the concrete floor in place while it develops its full strength, which takes several weeks. As each floor goes up, surveyors check the building is not getting out of true. The concrete floors in steel-framed buildings are created in much the same way.

The final stage in creating a skyscraper is to put into place the outer panels, which will form the walls. Because they bear no structural loads, they can be made of a wide variety of materials, including stone, brick, aluminium, stainless steel, tile, glass or concrete. The panels are made in factories and delivered to the site on trucks, hoisted into place and attached with bolts or other fasteners to the building's frame.

Old buildings can be given a "face-lift" by removing all the original panels and replacing them with something more fashionable, at a fraction of the cost of rebuilding the entire structure. Glass panels, which may be tinted or have a mirror-like surface, require special handling. The huge pieces of glass, up to an inch thick, are lifted with special suction cups to avoid damaging the edges as they are put into place.

The older buildings of Manhattan, like the Chrysler and Empire State buildings, are built of immensely strong steel and are very rigid, able to resist the force of the wind. Tests show that even in a high wind the Empire State Building bends only very slightly, less than $\frac{1}{4}$ inch at the 85th floor.

More modern buildings have less substantial steelwork, in order to reduce costs, so may need sophisticated arrangements to avoid sway. One example is at the Citicorp Building on Lexington Avenue, where a special damper in the form of a 400-ton block of concrete has been built into the 59th floor. The block is connected to the frame of the building by shock-absorber arms, and can be "floated" on a thin film of oil. When the wind blows strongly, oil is pumped under the block to lift it and allow movement. But because of the block's huge inertia, it moves only slowly and, through its links with the frame, stops that from moving also.

Beneath the ground lie the hidden but vital foundations upon which the stability of the

Erectors and Riveters

A head for heights was the obvious requirement for men working around a skeleton of girders hundreds of feet above the ground. A worker relaxes (below) while working on the Chrysler Building in 1928.

Until welding of steel girders became the glue of skyscrapers, the racket of riveting reverberated around Manhattan. Steel erectors would guide the girders into position, where they were secured by a riveting gang of 5 men. A "punk" supplied the "heater" with rivets to be heated on a forge. Glowing rivets were hurled to the "catcher" who collected them in a bucket, tapped each to remove cinders and rammed it into a hole for the "bucker-up" to hold while the "driver" flattened the end with a compressed-air hammer.

A steel erector uses the quickest way to reach his work on the Empire State Building in 1930. The Chrysler Building is in the background. A gang might put in 800 rivets during a $7\frac{1}{2}$-hour shift. Up to 38 gangs worked on the Empire State Building, at a time when no ear protection was available.

Erection of the Empire State Building steelwork (right), of 102 storeys, took just 6 months. As the Waldorf-Astoria Hotel was being demolished to make way for the Empire State Building, the stock market crashed, indirectly reducing the developer's costs: of the construction estimate of $44 million, about $20 million was saved.

An Iroquois steel erector (right) in front of the Chrysler Building in 1962. The Iroquois, who once inhabited New York State, are not the only workers of Indian ancestry to show a particular aptitude for work amongst the high girders; the Mohawk Indians from a reservation near Montreal have been at work pushing up the Manhattan skyline since the 1920s.

The Steel and Concrete Forest

building depends. The World Trade Center, built between 1966 and 1971 and briefly the tallest building in the world, has some of the most remarkable foundations ever built. An area equal to 16 football fields was dug out to a depth of six storeys, all below the level of the Hudson River.

To make this possible, a trench had to be excavated around the outside down to the depth of the bedrock, and filled with concrete to create a huge coffer dam. Then 1.2 million cubic yards of soil were removed to create the foundations for two towers 110 storeys high, rising to 1,350 feet. The soil removed was used to create 23 acres of new land on the banks of the Hudson River at Battery Park City next to the Trade Center. The twin towers at the Trade Center were briefly the tallest structures in the world, before being overtaken by the Sears Tower in Chicago, which is 1,454 feet high.

If New York property millionaire Donald Trump has his way, the title of the world's tallest building will return to New York. He has plans for a massive complex called Television City, on a 75-acre site on Manhattan's West Side. Trump bought the site, an abandoned rail yard, for $95 million in 1984, one of the best bargains since Dutch settlers bought Manhattan from the Indians who originally lived there for $24 worth of trinkets. On this site Trump intends to build six 70-storey towers (and one of 65 storeys) surrounding a central spire 150 storeys high, 216 feet taller than the Sears Tower. The project includes huge TV studios, apartment buildings, shopping malls and parks. It is the biggest project in Manhattan since the Rockefeller Center went up in the 1930s. "The world's greatest city deserves the world's greatest building", says Trump. "This is to be a great monument, majestic."

Trump has already built one 68-storey Trump Tower in Manhattan, on 5th Avenue next door to Tiffany's. It has a huge atrium, soaring upward for six floors, down which pours a constant stream of water in the world's highest indoor waterfall. The lobby gleams with acres of rose-red marble, there are smart boutiques selling expensive clothes and jewellry, and those who live in the apartments above enjoy magnificent views of Central Park. It is a flashy and highly fashionable building, and there seem no technical reasons to suppose that Trump cannot also carry off his even more ambitious vision in Television City.

"Donjons of a New Feudalism"

Central Park Lake in 1909 (below) and almost the same view in 1934 (bottom). In the centre of the later view is the Chrysler Building. The tall building with the gabled roof, centre left, is New India House.

So the *Illustrated London News* described the forest of skyscrapers that had grown up in Manhattan by 1934, when it presented an extraordinary comparison between the New York skyline of that year and just 25 years earlier. In 1909 the first buses were replacing horse-drawn carriages, and the fire service was still not mechanized; by 1934 the Empire State Building was dominating the Manhattan skyline and Lindbergh's solo flight across the Atlantic was already a memory 7 years old.

The Flatiron Building
(left) was built with 6
Otis hydraulic elevators.
It was Elisha Otis's
steam-powered "safety
hoist" that enabled
architects to disregard
the previous constraint
of a reliance on stairs.
Otis's breakthrough
was the invention of a
good braking system.
Elevators in the
Rockefeller Center
travel nearly 2 million
miles a year.

Continuous rebuilding
in Manhattan (above) is
not a new phenomenon:
when the Flatiron
Building was being
erected in 1901, the
nearby Hotel Pabst had
been demolished after
only 4 years. This view
looking south east
down 42nd Street shows
the Pan-Am Building
beyond the construction
site and the Chrysler
Building to the right of
centre.

The Geodesic Golfball

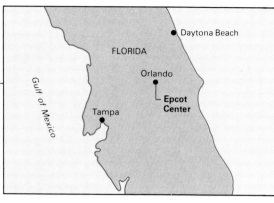

Fact file

The largest privately financed construction project ever undertaken

Builder: Walt Disney World

Built: 1966–82

Area: 260 acres

Walt Disney, the creator of Mickey Mouse and Donald Duck, had a dream. It was to create a futuristic city somewhere in America, complete with homes, schools, parks and jobs for all, where harmony would reign amid an environment planned by the best designers and engineers. That, at least, was what he sketched out on his table napkin some time in 1959. He even had a name for the people who would make this dream come true: he called them "imagineers".

What he got—or more accurately, what the corporate imagineers created 16 years after his death—was a theme park, set in 260 acres of land south west of Orlando in Florida. Epcot, originally supposed to stand for Experimental Prototype Community of Tomorrow, is not a community at all. Nobody lives there permanently, and at night it dies. By day, it is a thriving entertainment centre that attempts to educate while it is providing fun.

The building of Epcot was an extraordinary achievement, the largest privately financed construction project ever undertaken. Groundbreaking on the site took place in October 1979; it was opened on time on 1 October 1982. It cost $1 billion, double the estimate, and employed 600 Disney designers and engineers, 1,200 consultants, and 5,000 construction workers. To help pay for it, Disney executives went out into the world of commerce and persuaded a number of blue-chip companies to provide sponsorship. They ended up with seven, who put up $300 million over ten years. In exchange, they had their names attached to a pavilion apiece.

To an ordinary eye, the site Disney chose was not an appealing one; much of it was swamp, filled with muck, a peatlike organic material with a 95 percent water content. Drilling showed that in places this muck was 160 feet deep. Before the land could even take buildings and roads, it had to be either removed or consolidated. A total of 2.5 million cubic yards was removed, and replaced by 5 million cubic yards of clean material, while in the swampier areas the cushion of muck was simply compressed by 15 feet, covered with a blanket of sand, and turned into lagoons and lakes, with water up to 10 feet deep on top of it.

On this reconstituted ground, the Disney engineers erected buildings that house exhibitions, movies, restaurants, rides and instant townships representing the cultures of nine nations, including Britain, France, Italy and China. The theme is set by the most striking building of all, Spaceship Earth, a huge golfball that is the first completely spherical geodesic dome ever built. It is also the biggest, standing 180 feet, or 18 storeys, tall. It is made of steel framing clad with faceted aluminium panels, and stands on three pairs of steel legs.

Inside the huge sphere there is a spiralling ride that takes place in darkness, so that in fact one might as well be inside a boring square building. Its theme is communication, and it is sponsored by AT&T. The $\frac{1}{4}$-mile spiral ride passes various key events in human history—Cro Magnon Man painting the walls of his cave, Greek dramatists declaiming, Michelangelo labouring on the ceiling of the Sistine Chapel, Gutenberg manipulating type, and so on. The Disney publicity states that the hieroglyphics are authentic, the ancient dialects correct, and the costumes for the 65 animated figures have been exhaustively researched.

At the very top of the ride, there is a realistic simulation of floating in outer space before the descent. Forty thousand years of human history are condensed into 15 minutes, and the ride is so loaded with technological wizardry that, to begin with at least, it frequently went wrong. "We're sorry for the delay. Our journey in time has stopped for the time being", disembodied

The Geodesic Golfball

voices declared as the vehicles ground to a halt. "I have seen the future and it kept breaking down", one visitor concluded.

The hidden services of Epcot are to many people its most interesting feature. It has, for example, a fibre-optic telecommunication system (one of the first installed anywhere in the world), a pneumatic rubbish disposal system, PeopleMovers that are driven by linear induction motors, and an all-electric monorail that takes people to and from the neighbouring Disney World complex. (Monorails have been appearing in cities of the future for at least the past 25 years, though they seem no nearer to becoming part of the real world.) The waste at Epcot is used as fuel, to provide air conditioning and cook the food the visitors eat. There is a central security system that monitors 4,000 critical points for fire or distress and, as in all Disney theme parks, the cleanliness is extraordinary. Litter is scarcely allowed to hit the ground before it is whisked away and consigned to the pneumatic disposal systems, a possible model for the city of the future.

Epcot is divided into two sections, Future World (which includes the Spaceship Earth geodesic dome) and World Showcase. Future World consists of eight separate pavilions, including the Universe of Energy (courtesy of Exxon), where visitors are transported around in vehicles each accommodating 96 people. The electricity is supplied by 80,000 solar cells, providing 70 kilowatts, and the vehicles are directed by wires embedded in the floors of each room. There are two theatres, where the vehicles are rotated into position by turntables that are suspended on air, so that the occupants can experience multiscreen entertainments, or models of Mesozoic creatures lumbering around with the aid of silicon chips and solenoids.

A second pavilion in this section, The Land, is sponsored by Kraft. This time the tour is made by boat, and is accompanied by sounds, smells and hot winds: the chickens not only look like chickens, they smell like them too. It includes advanced agricultural techniques, such as a conveyor belt with lettuces growing in mid-air, their roots sprayed with water, subterranean irrigation systems, and fish farming.

The World Showcase is quite different, rather less like a vision of the future and more like a traditional Disney theme park. The pavilions here attempt the familiar idea of many a World's Fair, a "condensation" of the culture of many

Spaceship Earth (above) has been built to withstand wind speeds of up to 200 mph— Florida is prone to hurricanes. The perfectly spherical geodesic dome rests on 6 steel legs 30 feet wide that raise it 15 feet above the ground. The interior of one of the 17 major pavilions, "Journey Into Imagination" (left), explores the latest ways of producing images, and creates dramatic effects for visitors.

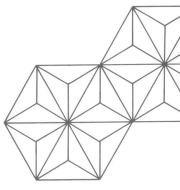

The framework of the geodesic dome consists of 1,450 steel beams covered with waterproof neoprene sheeting. The external cladding of nearly 1,000 triangular aluminium panels (left and above) is bolted on to the steel framework. The sphere has a diameter of 165 feet and a volume of 2.2 million cubic feet, but the darkness prevents visitors gaining any sense of size.

The monorail (below) has become a feature of most World Fairs and futuristic representations, but of few cities. The existing monorail system to transport guests around Walt Disney World was extended by 7 miles to take in Epcot, much of it on an elevated beamway designed for speeds up to 45 mph. The beamway is of precast, prestressed, post-tensioned concrete with spans of 120 feet.

nations complete with the olde English pub, a Chinese garden, a souk from Morocco. Here the illusionists' art perfected by generations of successful cartooning is seen at its best. None of the buildings is made of the proper materials—glassfibre predominates—but it is all put together with enormous skill. Even the flaws that a real building might have, such as a chip in the stucco, have been faithfully recreated.

"Of course it is hokey" admitted one architectural critic, impressed despite himself. "But it has been so carefully considered and so expertly executed by real artisans (Disney raised the mastery of glassfibre to craft, and an art director accompanied each construction crew) that it bears the signature of the skilled human hand in design and execution."

The Glass Tent

Fact file

One of the world's most extraordinary roof structures covering Europe's largest hall

Architect: Guenther Behnisch and Partners

Built: 1966–72

Materials: Concrete and glass

Roof area: 89,700 square yards

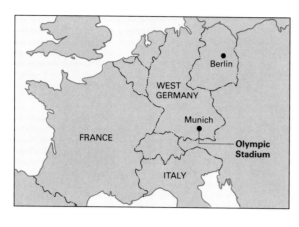

The stadium's roof at sunset resembles a Bedouin tent. The 58 masts that hold up the roof are so positioned that the pressure on the glass and on the frame in which the panes sit is equalized.

Creating the right environment for the Olympic Games has become a major challenge and a huge expense for the city that wins the right to stage them. Few have met this challenge more successfully than Munich, the Bavarian city that was the site for the games of 1972. The eye-catching stadium designed for those games by the Stuttgart architects Guenther Behnisch and Partners has never been equalled. Even today, almost 20 years since it was built, it remains one of the most remarkable structures in Europe.

The feature that distinguished the Munich stadium was its extraordinary roof, which consists of 89,700 square yards of acrylic glass plates, supported in a tentlike structure by huge masts and cables. It looks like a cobweb, too delicate to survive the first storm, but that is an illusion. The stadium is still in full use today, having staged more than 3,500 sporting, cultural and commercial events attended by more than 30 million visitors since the Olympic Games ended.

The Olympic Park, as it is called, is in fact a lot more than just a graceful roof. In addition to the stadium, it includes two halls, an indoor swimming pool, an ice-rink, a cycling stadium, a lake, and a tower 941 feet high which is one of the highest buildings in Europe. The roof flows across a large part of the park, providing cover for both of the stadiums, the Olympic Hall, the swimming pool and the pedestrian areas.

Planning began in 1966 when the International Olympics Committee awarded the games of the XXth Olympiad to Munich. The site was on the former army training grounds of the kings of Bavaria, and the adjoining Oberwiesenfeld airfield, a total of 225 acres.

The first element to be finished was the Olympic Tower, which had been planned before the city won the games. Built of reinforced concrete with two viewing platforms and a restaurant at the top, the tower was completed in 1968. It is used for TV broadcasting, as well as proving a considerable tourist attraction. There are two elevators, operating at 22 feet per second, to carry the two million visitors a year up to the viewing platforms or the 230-seater restaurant, which revolves through 360 degrees in 36, 53 or 70 minutes, depending on the speed setting. The entire tower weighs 52,000 tons.

The view from the tower includes the whole Olympic Park, the city of Munich and in the distance, the Alps. It provides a spectacular perspective on the roof which billows like canvas below. The structure is supported by 58 masts, which in turn are fastened to anchors in the ground by steel guy ropes. The masts are all placed on the periphery of the roof, so that inside the stadium there are no obstacles to a perfect view. From the masts further cables support the roof itself, which rises to a series of points where the cables are attached.

The roof consists of a series of acrylic glass plates, up to 9 feet 9 inches square and just $\frac{1}{6}$ inch thick. The glass is intended to be self-cleaning, since crawling about on the roof is not an easy task. It depends on the rain, snow and frost to remove dust and dirt that settles on it; fortunately, Munich is not heavily industrialized.

Each of the plates is surrounded by light metal rails, into which they are sealed by neoprene rubber buffers to make the system waterproof, and to allow for movements that can occur as a result of changes in temperature or during storms. The light metal framework does not, however, actually support the weight of the roof. That is done by a second network of cables that run across the roof in both directions at a spacing of 30 inches and are attached directly to the glass plates, not to the framework.

The connection between the supporting network and the glass is made by a series of steel bolts about 4 inches long attached to the plates.

The Glass Tent

The joint is once again buffered with a neoprene washer, to equalize pressure and to take up any shocks. In total, there are 137,000 such joints over the entire roof. Effectively the roof thus consists of a network of cables running in two directions at a 30-inch spacing to produce a grid, supported from above by the masts and anchored to the ground at various points. From this web hang the glass plates. The arrangement is designed so that pressure is equalized over the surface of the glass, and no pressure is exerted on the frames in which the glass sits. This is intended to eliminate the possibility of one of the frames distorting and allowing a plate to fall.

The roof covers just over half of the Olympic Stadium, with the other half unprotected. The stadium is built of reinforced concrete and at its highest point the west stand rises to a height of 110 feet. The capacity of the stadium is 78,000, of whom 48,000 are seated. The football pitch, on which West Germany won the 1974 World Cup final, is heated by about 12 miles of plastic piping under the turf, which ensures that the pitch is playable even in the worst conditions. The stadium has also been used for concerts by pop groups, and it was here that the Pope blessed the Catholics of Germany. The World Congress of Jehovah's Witnesses has also been held here.

The Olympic Hall, used for handball and gymnastics in the 1972 games, is a completely covered arena with seating for 14,000 people, a capacity unmatched anywhere else in Europe. The hall is 585 feet long, 390 feet wide and 136 feet high. The tent roof covers it, while the walls consist of a glass façade up to 60 feet high. The Olympic Hall has staged six-day cycle races, Davis Cup tennis matches, the World Ice Hockey championships, performances of *Aïda* and concerts by Tina Turner and Luciano Pavarotti. It is one of the largest and most flexible spaces anywhere, able to host a schools sports gala one day, a pedigree dog show the next, and moto-cross races the day after.

The Olympic Park also boasts one of the finest swimming-pool complexes in Europe, with five separate pools—designed for competition, diving, training, practice and a special pool for children. Like the Olympic Hall, the pools are covered by the roof, with a glass façade rising to more than 80 feet in places providing the walls. This creates the effect of an outdoor pool, even though it is protected from the weather and can seat 2,000 people. During the games, temporary seating was provided to more than quadruple the

The adaptability of the roof structure is evident in this view from the Olympic Tower. The system enables large and small halls and stadia to be linked by a continuous ceiling of glass, which gives the site a cohesion and unity that separate units could not achieve.

capacity, bringing it up to 9,360.

Providing a site for the Olympics has proved a good investment for Munich. The total cost of landscaping the site, and building the many sports facilities, cost DM 1,350 million, two thirds of which was covered by revenue raised by the organizing committee, the sale of Olympic coins, a TV lottery and a public lottery. Half the rest was financed by the Federal Republic government, the other half jointly by the Bavarian State and by the city of Munich. Given a repayment term of six years, the city was able to finance its share out of the current budget each year—and finished with a sports complex any city would envy. It is a tragedy that the games for which the complex was built will be remembered principally for a terrorist attack on Israeli athletes which overshadowed the performances on track and field; but the Olympic Park has proved a great success in the years since then.

The Olympic Tower dominates the Munich skyline and affords the best view of the Olympic Park. Erected in 1962, before most of the buildings that form the Park, the Tower is 941 feet high and weighs 51,200 tons. Elevators ascend the tower until 11.30 pm, giving visitors $\frac{1}{2}$ hour to enjoy the lights before the last ride down.

Deterioration of the neoprene joints (above) will eventually necessitate their complete replacement.

The Stadium (left) accommodates 52,448 seated spectators and 20,608 standing, and there are 240 seats for press, television and radio reporters with 20 commentators' cabins. The West Stand is built over the athletes' and technical facilities. The 2 indicator boards are so large that 48,000 light bulbs are used.

Australia's Architectural Symbol

AUSTRALIA

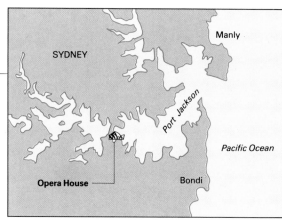

SYDNEY

Manly

Port Jackson

Pacific Ocean

Opera House

Bondi

Fact file

One of the most distinctive buildings in the world, that required novel construction ideas

Designer: Jorn Utzon

Built: 1959–73

Materials: Pre-stressed concrete and glass

Weight of roof: 26,800 tons

Area of glass: 1½ acres

Sydney Opera House has become an Australian symbol as instantly recognizable as the kangaroo or the koala. Its gleaming white roofs cluster like overlapping shells on a promontory that juts into Sydney Cove, creating a building that looks beautiful from any angle.

Beneath the shells are five separate halls—for symphony concerts, operas, chamber music and plays—as well as an exhibition hall, 3 restaurants, 6 bars, a library and 60 dressing rooms. The Opera House has a thousand rooms and 11 acres of usable floor space. A million ceramic tiles cover the roof, and 67,000 square feet of specially made glass fill the open ends of the shells. Behind the creation of this extraordinary building lies the inspiration of one architect and the work of many, combined with the skills of the structural engineers who made it possible. There is no other building like it, anywhere. It is unlikely there ever will be.

The design was the prize-winning entry in a competition launched in 1955 by the Premier of New South Wales, Joseph Cahill, to create a

national opera house on a magnificent site in Sydney Harbour. Bennelong Point is named after an Aborigine befriended by the commander of the First Fleet, Captain Arthur Phillip RN, who landed with Australia's first convict settlers in Sydney Cove in 1788. From 1902 it was occupied by a huge red-brick tram depot, which was demolished to make way for the Opera House.

To the astonishment of the world, the contest was won by a little-known Danish architect, 38-year-old Jorn Utzon, who had few completed commissions to his credit. Almost all he had built were 63 houses near Elsinore in 1956, and a smaller housing project near Fredensborg. For the Opera House, Utzon had produced a design so graceful and daring that it simply swept the others aside. He provided few details. "The drawings submitted are simple to the point of being diagrammatic" the judges concluded. "Nevertheless, as we have returned again and again to the study of these drawings, we are convinced that they present a concept of an opera house which is capable of being one of the great buildings of the world."

Faced with this wonderful but difficult design, the New South Wales Government might have lost its courage; it was under no obligation to build the first prize winner. It could have saved itself a lot of money, and years of argument, if it had chosen a simpler but more ordinary structure. But it did not. It accepted Utzon's design and at his suggestion appointed Ove Arup and Partners, the British-based firm founded by a Danish engineer, as structural consultants.

The first stage was to clear the site and build the podium, the deep, flat platform upon which the building stands. Work on that began in 1959, before it was clear that Utzon's shells could even be built. The assumption at this stage was that the roofs would be made of a vault of concrete poured in one operation into curved wooden or steel moulds. That would, however, have been prohibitively expensive, so Utzon came up with another idea.

He suggested instead that the shells should be made of prefabricated concrete ribs, standing next to one another, and that they should all have the same spherical curvature. He showed that all the shells could be made from sections taken from the surface of a sphere 246 feet in radius, like pieces cut from the skin of an orange. But instead of casting the shells as one piece, they would be made up of ribs, cast in separate sections on site from a relatively small number of

The two principal halls (left)—for opera on the left and orchestral concerts on the right—appear joined from most angles; in fact, like the small set of shells covering the restaurant (right), they are divided by a passageway. The buildings cover 4½ acres of the site area of 5½ acres, and the 5 auditoria provide seats for 5,467.

Australia's Architectural Symbol

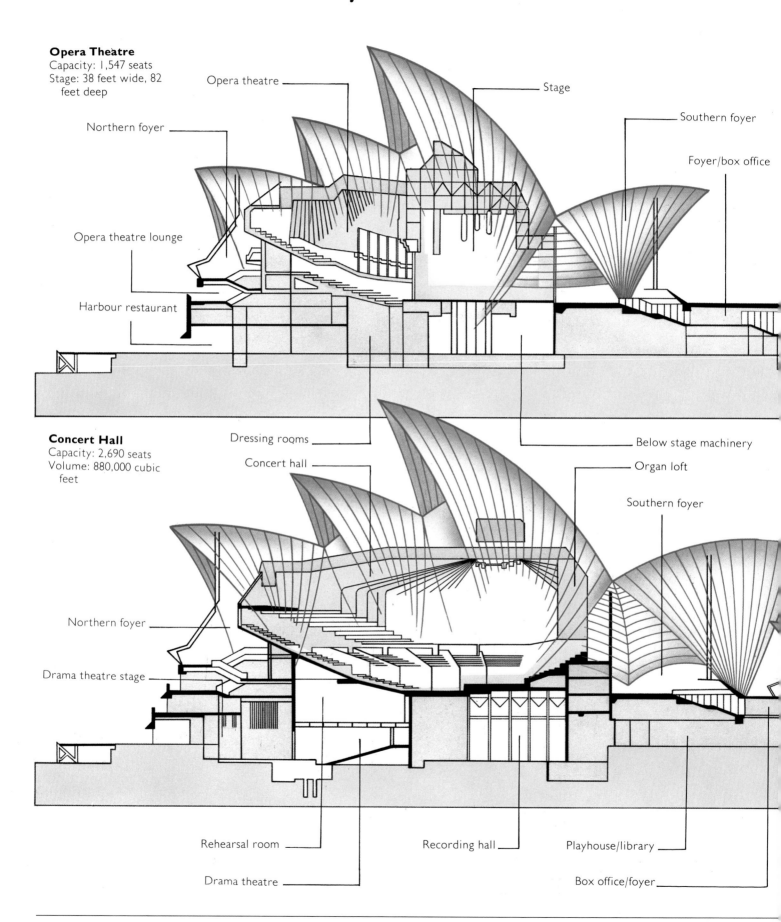

Opera Theatre
Capacity: 1,547 seats
Stage: 38 feet wide, 82
feet deep

Opera theatre

Stage

Northern foyer

Southern foyer

Foyer/box office

Opera theatre lounge

Harbour restaurant

Concert Hall
Capacity: 2,690 seats
Volume: 880,000 cubic
feet

Dressing rooms

Concert hall

Below stage machinery

Organ loft

Southern foyer

Northern foyer

Drama theatre stage

Rehearsal room

Recording hall

Playhouse/library

Drama theatre

Box office/foyer

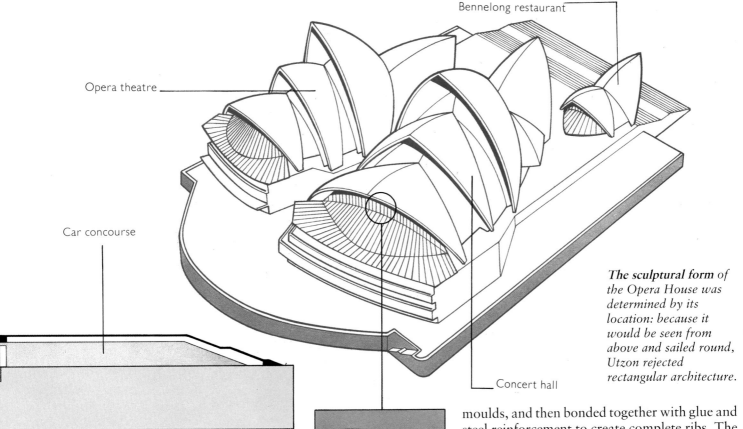

Bennelong restaurant

Opera theatre

Car concourse

Concert hall

The sculptural form of the Opera House was determined by its location: because it would be seen from above and sailed round, Utzon rejected rectangular architecture.

Bennelong restaurant

Car concourse

Staircase to foyer

The choice of glass to fill the mouths of the shells was made by Utzon at an early stage, but the technical problems of supporting such a vast area, and of keeping out the noise of shipping sirens, were daunting. Utzon was determined to avoid a vertical wall since it would "kill the unsupported shell effect"; consequently he opted for a broken line.

moulds, and then bonded together with glue and steel reinforcement to create complete ribs. The ribs stand so close together they almost touch, and are linked to one another by a concrete joint. The outside of the shells is then covered by a layer of ceramic tiles.

Selecting and placing the tiles was a huge problem. Utzon regarded the choice of roof covering as vital: "The wrong material would ruin the appearance", he wrote. It would have to gleam in the sun, survive large temperature variations, keep itself clean, and retain its character for many years. To find such a material, Utzon looked to the ancient world, and decided that ceramic tiles were the only answer. The use of the spherical curves for the ribs meant that the surface could be covered successfully with tiles of a single size, $4\frac{3}{4}$ inches square.

Two different surface finishes were used, one glossy and white, the other matt and buff in colour, to give the roof its distinctive pattern. The tiles, made by Hoganas in Sweden, were laid on the roof in prefabricated trays, or "lids". These were assembled by laying the tiles face downward on a form and pouring concrete over their backs to bond them together. The complete lid, which could be as much as 33 feet long by 7 feet 6 inches broad, was then removed from the form and bolted to the roof by phosphor-bronze bolts. The entire roof contains 4,253 lids, made up of 1,056,000 individual tiles.

One of the mysteries of the building is how the

Bennelong Point is one of the finest situations chosen for a public building, backed by the green acres of the Royal Botanic Gardens and Government House grounds. In April 1964, when the P&O liner Canberra *passed the site, work was beginning on the halls.*

shells are supported, apparently on only two points, without the use of pillars. This is achieved by linking the large shells to smaller ones facing in the opposite direction, so that the two form a unit. Since each shell touches the ground at two points, this means that the unit is squarely set on four legs. The smaller "louvre shells", as they are called, are barely noticeable, but without them the roof could not stand up.

Another tricky problem was devising a system for glazing the open ends of the shells. Utzon always intended to use glass for most of the shells, but finding a way to support it was difficult. The glass walls were eventually supported on vertical steel mullions which extend all the way up the mouths of the shells. From these mullions run bronze glazing bars into which the panes are embedded in silicone putty. There are 2,000 panes of glass, ranging in size from 4 feet square to 14 feet by 8 feet 6 inches, in more than 700 different sizes, worked out by Ove Arup and Partners using a computer. The glass itself is $\frac{3}{4}$ inch thick and consists of two layers, one plain and one amber, joined together by an interlayer of plastic. This structure increases the strength of the windows, reducing the danger from falling glass inside the building, and also produces better sound insulation.

By 1966, work on the main structure was well advanced, but little had been done inside the building. Utzon had fallen out with government officials over methods of construction and the awarding of subcontracts. A new state government, elected to office in May 1965, was becom-

ing anxious about the costs of the building, already certain to exceed the original estimates by a wide margin. Suddenly, in February 1966, Utzon resigned as architect to the Opera House. Despite appeals from the government to return to the project as a member of the architectural team, not its supervisor, he stuck to his decision. His letter of resignation contained the gnomic sentence: "It is not I but the Sydney Opera House that created all the enormous difficulties." Soon after he left Australia, never to return.

His task was finished by a team of Australian architects, so that while the exterior of the building bears Utzon's stamp, the interior does not. Decisions were also taken which changed

*The **partly completed shells** in June 1965 illustrate the prefabrication of the roofs, which are made of 2 rows of concrete ribs that curve inwards to meet at the centreline. The roofs required 2,194 precast segments, which were made in a casting yard on the site.*

The tiling of the roof posed huge difficulties and required surveying techniques using computers to resolve them. The impossibility of using regularly shaped tiles on Utzon's original shells, whose contours altered continuously, forced a change in shell design.

the nature of the building, committing the largest hall (which seats 2,690) not to opera but to orchestral music. Utzon had believed that it could serve both purposes, but a panel set up to advise the government after his departure recommended otherwise. As a result, the Sydney Opera House is unable to stage the grandest of operas, which require complex stage machinery and a large orchestra pit. The hall used for opera, which seats 1,547, was originally intended for plays, and does not have a pit large enough. As a result, says conductor Sir Charles Mackerras, "it's very nearly impossible to do anything adequately". When the Australian Opera puts on Wagner's Ring, they have to use a reduced version. Though large-scale operas are put on in the bigger hall, it lacks proper stage machinery.

Because of these shortcomings, opera lovers have always regarded the house as something of a fraud: an opera house containing a large concert hall not suitable for opera and a smaller hall satisfactory only for small-scale works. Defenders of the decisions taken after Utzon's departure argue that the term Opera House has

always been a misnomer anyway, because the competition rules made it clear that opera was not the building's primary function.

Despite the arguments, the building was eventually completed, and opened by Queen Elizabeth II in October 1973. The original estimate given by Joseph Cahill had been A$7 million (£3.5 million); the final cost came to A$102 million (£50 million), the vast majority raised by lotteries. The state government breathed a sigh of relief and began to enjoy the international praise for its stunning new building.

However, by March 1989, Parliament was warned that urgent repairs were needed, at a cost of A$86 million (£42 million), if the building were not to be irretrievably damaged. Tiles had begun falling off the roof, which was leaking, as were some windows and walls. Sealants used on the concrete ribs, expected to last for 20 years, had deteriorated after only ten. Whatever the cost, the building would be maintained at the highest standard, the New South Wales Arts Minister told Parliament, but the Opera House is clearly going to remain demanding for years to come.

The ceiling of the Concert Hall is designed to create a space with acoustic properties suitable for music and speech. A hollow raft composed of layers of concrete, plasterboard and plywood is suspended from the ceiling. The cavity conceals and allows access to wiring and air-conditioning ducts. The floor is of laminated brush box.

The Ultimate Stadium

The world's largest indoor stadium, in the centre of New Orleans, is a colossal multi-purpose building with the biggest dome ever constructed. The roof alone covers almost 10 acres, rising in the centre to the height of a 26-storey building. Halfway through building it, the subcontractor entrusted with the job walked out, claiming that the design simply would not work. The architects who designed it kept calm, found a new subcontractor and finished the job. More than 15 years later the building is a huge success which has vindicated the design and the decision of the Louisiana Stadium and Exposition District to build it.

The Louisiana Superdome is not the only enclosed stadium suitable for sports events, rock concerts and political conventions in the US, but it is the biggest. Its dome is 680 feet across and 273 feet high in the centre, and its seating can accommodate up to 75,000 people for sports events, or more for special events such as concerts.

Planning began in 1967 when LSED announced its intention of creating such a building in a derelict part of downtown New Orleans, then occupied by rusting railway tracks and abandoned warehouses. The contract was won by the New York firm of Curtis and Davis Architects and Planners, who formed a joint venture with the St Louis engineering consultants Sverdrup & Parcel and Associates. The contract was awarded early in 1971, and the first concrete pile was driven into the soft Louisiana mud in August 1972.

The key to the economic success of the building was flexibility, so that it could be used for a range of sports—American football, baseball, basketball, and even tennis—as well as conventions, trade shows, theatrical productions, and large-scale closed circuit TV events. The architects sought to produce a building "large enough to house the most spectacular extravaganza and small enough to accommodate a poetry reading", in the words of the leader of the design team, Nathaniel C. Curtis Jnr. For this reason, a number of meeting rooms and convention facilities were placed immediately behind the main stadium seating.

Satisfying the needs of a variety of sports is more difficult, for baseball needs the biggest field, while football attracts the most spectators, and basketball and tennis are usually played in smaller arenas where the atmosphere is more intimate. To meet these different requirements,

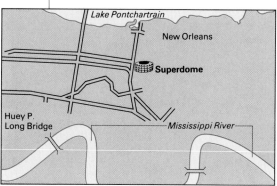

USA

Lake Pontchartrain
New Orleans
Superdome
Huey P.
Long Bridge
Mississippi River

Fact file

The world's largest indoor stadium covered by the world's biggest dome

Designers: Curtis and Davis

Built: 1971–75

Materials: Steel and concrete

Width of dome: 680 feet

15,000 seats along the lower concourse can be moved to and fro. In their rear expanded position, they provide the maximum area for playing baseball, but for football they can be moved forward 50 feet to give spectators the ideal view close to the touchline. For tennis, an entire section of the east stand, with 2,500 seats, can be moved right across the field to form a compact arena on the west side of the stadium.

The building is supported on 2,100 prestressed concrete piles driven 160 feet down to bedrock. At the lowest level are three tiers of parking for 5,000 cars. Above that is a floor of offices, and above that again the convention floor. At a height of 160 feet above the ground, a circular ring of welded steel supports the domed roof. This tension ring, 680 feet in diameter, is in the form of a truss 9 feet deep and 18 inches across. Designed to withstand the immense forces exerted by the dome, the tension ring is made from 1½-inch steel, prefabricated in 24 sections that were welded together in place. To ensure the reliability of the welds, they were carried out in the controlled atmosphere of a tent-house that was moved around the circumference of the ring from joint to joint.

The dome itself consists of a framework of structural steel radiating outward from a central "crown block". During construction, the dome was supported by 37 temporary towers, each

The Superdome's design was consciously intended to facilitate the range of activities that were held in the amphitheatres of ancient Greece, from sport to poetry readings. Determining the optimun seating arrangements around the different-shaped fields for football and baseball (right) required the computer analysis of 200 schemes.

with a hydraulic jack on top so that the entire dome, when complete, could be lowered on to the tension ring which was to support it. The dome's structure is formed by 12 curved radial ribs running outward from the centre, linked by six sets of circumferential ribs and braced by a series of other struts to create a diamond pattern. The weatherproof surface on top consists of 18-gauge sheet steel panels, followed by a 1-inch layer of polyurethane foam and finally a sprayed-on layer of hypalon plastic. The idea was to create a homogeneous skin with no joints, and with enough flexibility to allow the structure to expand and contract by several inches in response to temperature changes.

For the same reason, the tension ring on which the dome rests is supported by pinned connections that can hinge to allow movement. Up to 8

The Ultimate Stadium

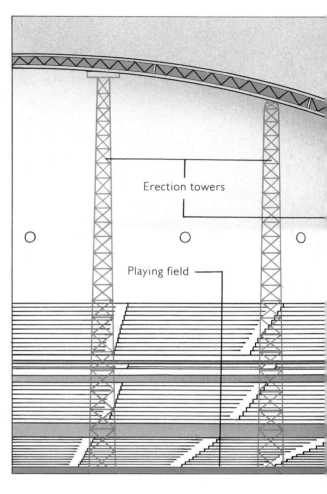

The Superdome's location in downtown New Orleans was chosen to take advantage of existing public transport and highway systems, and 20,000 commercial parking spaces nearby (left). The white plastic covering to the roof, designed to avoid joints, was sprayed on inside a cocoon for protection from the weather.

Erection towers

Playing field

The frequency of hurricane winds in New Orleans necessitated wind tunnel tests on a 1:288 scale model which proved that the Superdome could withstand sustained winds of over 150 mph and gusts up to 200 mph. The steel framework of the walls and roof required 20,000 tons of steel from Pittsburgh, delivered by barge up the Mississippi River.

inches of movement occurs in normal conditions. Between the dome and the outer wall is a trench 4 feet deep by 8 feet wide which forms a gutter to collect the rain falling on the roof. This trench, or gutter tub, can hold the equivalent of 1 inch of rain falling on the roof; to avoid overtaxing the sewers of New Orleans, the flow of water away from the gutter is controlled by drain pipes that do not allow it all to escape at once.

When the structure of the dome was complete, the hydraulic jacks on the top of each tower supporting it were removed, one by one, until it rested completely on the tension ring. The fabricators of all the steelwork in the Superdome, American Bridge, had estimated that when support was removed from the centre of the dome, it would settle by 4 inches under its own weight. They were delighted when the

actual settlement turned out to be $3\frac{1}{2}$ inches.

The circular shape of the domed roof creates an immense amount of lift as winds blow across it, acting rather like an aircraft wing. This is counteracted by the weight of the roof itself, assisted by a 75-ton TV gondola suspended inside the stadium from the centre of the dome. The gondola contains six giant TV screens, each 22 by 26 feet, sound systems, and lighting. A control room projects the TV pictures on to the screens from six projectors in the upper tier seats, and can provide instant replays, TV pictures from other stadia or events, or messages. The height of the gondola can be adjusted. For football it is placed about 100 feet above the field, but is raised to 200 feet for baseball to prevent stray balls hitting the screens. For theatrical events the gondola can be lowered to whatever height the director prefers, or the six screens replaced by two larger ones for showing sports events, motor racing or theatre productions from outside.

Given the very soft soil under the foundations, there was a theoretical danger that one or other

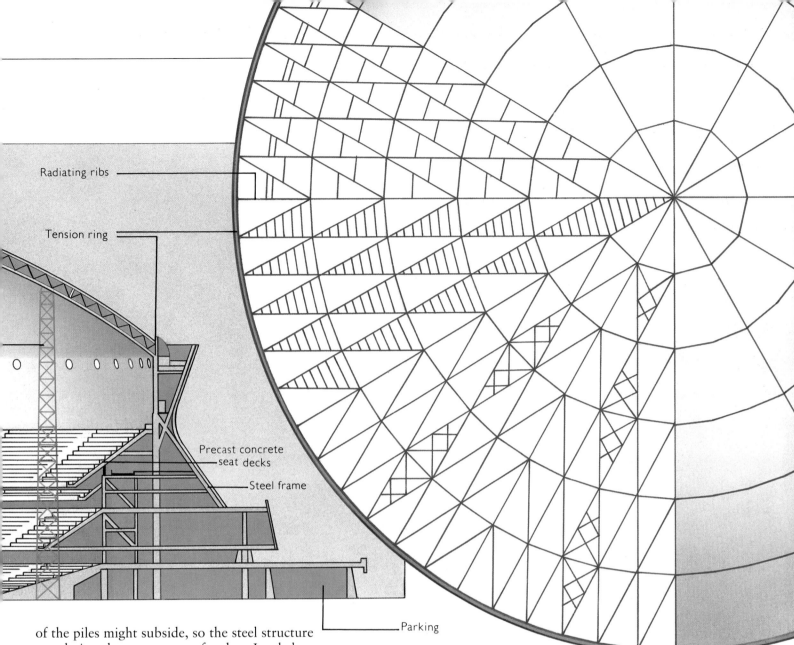

Radiating ribs

Tension ring

Precast concrete seat decks

Steel frame

Parking

of the piles might subside, so the steel structure was designed to compensate for that. Just below the tension ring very heavy steel cross-bracing was added to the structure, so that if a column were to sink, load would be redistributed to the two neighbouring columns. In theory, the bracing is strong enough to suspend completely a settling dome column, although in practice this will not happen. Settlement only happens under load, and as the bracing redistributes that load, settlement stops.

The Superdome first opened its doors on 3 August 1975. Since then it has been home several times to America's biggest football game, the Superbowl, housed the Republican National Convention in 1988, and an address by Pope John Paul II to 88,000 schoolchildren in 1987. It holds the record for the biggest indoor concert crowd in history (87,500 to hear the Rolling Stones in 1981) and has become a major tourist attraction in its own right, with some 75,000 people a year attending its daily tours. The grass on which football and baseball are played is artificial, AstroTurf 8—or Mardi Grass as the

Superdome operators call it. At first, the Superdome was managed by the State of Louisiana, but it lost money and was handed over to a private firm, Facility Management of Louisiana, which has achieved much better results.

The total cost of the building was $173 million, and in its first ten years of operation it cost an additional $99.2 million, for interest and repayment on the bonds sold to finance it, operating subsidies, and capital improvements. But a study by the University of New Orleans shows that the benefits to the area far exceed the costs. It estimates that new money brought to the area by the Superdome is almost $1 billion over that period, while the benefit to the rundown section of the city where it was built could not have been achieved in any other way. In 1970 the area was one of urban decay; today it is one of the showcases of the business district.

The 6-ring lamella roof-framing system was constructed on a patented configuration in which a diamond pattern is created by imposing parallel ribs on the 12 radial ribs and 6 rings. Each ring is made up of numerous small lattice trusses.

Symbol of a City

Fact file

The world's tallest unsupported structure

Builder: Canadian National Railways

Built: 1973–76

Material: Concrete

Height: 1,815 feet 5 inches

Weight: 130,000 tons

Cost: $57 million

The skyline of downtown Toronto is dominated by a pencil-slim tower that soars above the office blocks around it. The CN Tower, a structure of the television age, was built by Canadian National Railways for the practical purpose of eliminating "ghosting" that ruined the picture on many local screens. It has achieved much more than that. Today it is the symbol of Toronto's newfound confidence, the tallest unsupported structure in the world, visited by nearly 2 million tourists every year.

The tower rises 1,815 feet 5 inches from its base to the tip of its lightning rod. It weighs 130,000 tons, and its concrete and steel foundations rest on a bed of specially smoothed shale 50 feet below ground. Construction began on 6 February 1973 and took just 40 months, at a cost of $57 million. In cross-section the tower forms the configuration of a Y, which tapers gently as it rises. More than 1,100 feet up, a doughnut-shaped bulge is wrapped around the tower: it is a

seven-storey building, the Skypod, containing the equipment needed for TV broadcasting, a revolving restaurant, two observation decks, a night-club and two small picture theatres. Higher still, at 1,500 feet, is another even more vertiginous observation platform, the two-storey Space Deck, whose windows curve in toward the floor to enable the brave to look vertically downward, a spine-tingling experience; on a clear day one can see for 100 miles.

The tower was built by pouring high-quality concrete into a huge mould, called a slipform, supported by jacks which gradually inched it upward. As it climbed, the slipform was reduced in size to produce the taper of the tower. Because of the speed of construction—up to 20 feet a day—normal testing of the strength of the concrete (which involves waiting seven days for it to harden) could not be used, and special accelerated tests had to be adopted.

Extraordinary efforts were made to ensure the tower did not lean or develop a twist. In addition to a plumb bob (a 250-lb steel cylinder suspended on a wire down the centre of the tower's hexagonal core) optical instruments were used to take measurements every two hours. The result was a tower 1,815 feet and 5 inches high which deviates only $1\frac{1}{10}$ inches from a perfect vertical.

The top 335 feet of the tower consist of a steel transmission mast, assembled from 39 sections by lifting them into place one by one suspended from a huge Sikorsky S64E helicopter. This flying skyhook lifted and placed the sections in

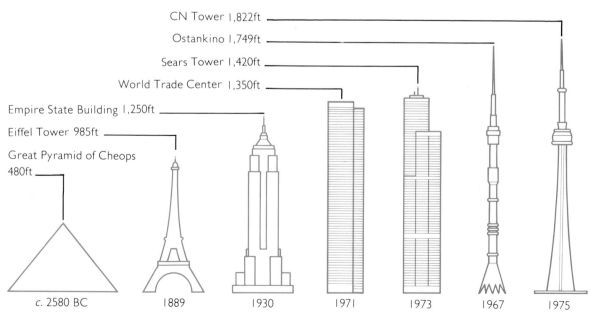

The height of the world's tallest unsupported building has almost doubled (left) in less than 100 years between the completion of the Eiffel and CN towers. The lighting of the CN Tower (right) has to be reduced during the spring and autumn bird migrations to avoid attracting them, thus risking fatal collisions.

Symbol of a City

three and a half weeks, instead of the six months it would have taken by more conventional methods.

The tower was designed to resist the worst weather imaginable, and then some. A wind of 130 mph is expected to occur once every thousand years; the tower was built to withstand one twice as strong. If you were bold enough to go up to the Skypod in a 130 mph wind, you would find yourself wobbling in an elliptical path through a distance of about 10 inches. But the movement would be so slow that it would be barely perceptible. The steel television mast would bend far more, oscillating through an 8-foot path, so special lead counterweights have been added to damp out the effect. The tower is a wonderful place to watch a thunderstorm, since it acts as a lightning conductor for all the surrounding buildings. Each year it is struck by lightning at least 60 times, the charge being harmlessly grounded to earth.

A greater danger with structures that rise as high as the CN Tower is the formation of ice high above the ground. Melting slabs of ice are reported to have fallen from the Ostankino Tower in Moscow, the second highest in the world, threatening the life of anyone below. In Toronto, the danger has been eliminated by ice-proofing areas where ice might form, such as the roof edges of the Skypod. In some places heating cables have been inserted, while in others a sheath of shiny plastic has been attached, on which ice cannot cling.

Visitors are carried up the tower in lifts which zoom upward at 1,200 feet a minute, as rapid a rate of climb as a jet taking off. The speed and acceleration of the lift were carefully calculated so that it is fast enough to be enjoyable, but not so fast as to cause alarm, nausea, or fainting fits. There is a glass wall to see the view, but the lifts have also been designed to provide a cocoonlike sense of security. The lifts have independent power supplies and can empty the tower quickly in an emergency, but to reassure visitors, there is also a staircase of 2,570 steps.

The tower has been the scene of many stunts. The first man to parachute off it was a member of the construction crew, Bill Eustace, also known as "Sweet William", on 9 November 1975. He was sacked. In 1979, the record for dropping an egg was smashed when Patrick Baillie, 17, threw a Grade A egg undamaged from 1,120 feet into a specially designed net. All this and better TV pictures too.

A Sikorsky S64E helicopter was used to locate the crane that was mounted on top of the Space Deck during construction. It was then used to position the 39 sections of the antenna that sits on top of the Space Deck; the heaviest section of the antenna weighs 8 tons.

Painting parts of the transmission mast (below) took 4 men 11 days, working 1,815 feet above the railroad tracks. The mast is protected from ice build-up by a glass-reinforced plastic covering 2 inches thick.

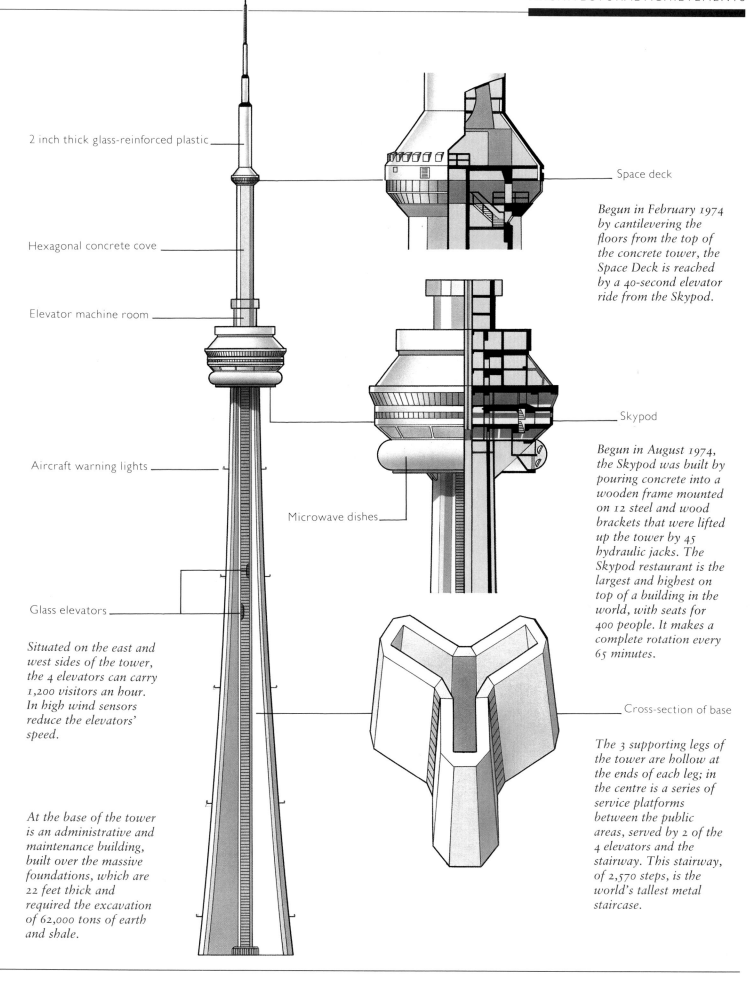

2 inch thick glass-reinforced plastic

Hexagonal concrete cove

Elevator machine room

Aircraft warning lights

Glass elevators

Space deck

Begun in February 1974 by cantilevering the floors from the top of the concrete tower, the Space Deck is reached by a 40-second elevator ride from the Skypod.

Skypod

Begun in August 1974, the Skypod was built by pouring concrete into a wooden frame mounted on 12 steel and wood brackets that were lifted up the tower by 45 hydraulic jacks. The Skypod restaurant is the largest and highest on top of a building in the world, with seats for 400 people. It makes a complete rotation every 65 minutes.

Microwave dishes

Situated on the east and west sides of the tower, the 4 elevators can carry 1,200 visitors an hour. In high wind sensors reduce the elevators' speed.

At the base of the tower is an administrative and maintenance building, built over the massive foundations, which are 22 feet thick and required the excavation of 62,000 tons of earth and shale.

Cross-section of base

The 3 supporting legs of the tower are hollow at the ends of each leg; in the centre is a series of service platforms between the public areas, served by 2 of the 4 elevators and the stairway. This stairway, of 2,570 steps, is the world's tallest metal staircase.

The Tallest Towers

One of the 7 wonders of the world was the Pharos of Alexandria, a tower built to help navigation off the Mediterranean port during the reign of Ptolemy II, who died in 247 BC. It is estimated to have been about 350 feet high, with a lantern section on top. It collapsed in 1326.

Successive ecclesiastical buildings have held the record for the tallest structure in the world until the Washington Memorial took the record for just 5 years in 1884. Since World War II, radio and television masts have held the record, the current holder being the Warszawa Radio mast near Plock in Poland which is 2,120 feet 8 inches high, or over $\frac{2}{5}$ mile. Plans for ever taller towers are continually made.

Bishop Rock Lighthouse, England

Standing between the Scilly Isles and the Lizard, Cornwall, this lighthouse was the first indication for many mariners that the transatlantic crossing was almost over, besides warning of submerged rocks. It is the tallest lighthouse in Britain, with a height of 156 feet 10 inches to the helipad, which is a recent addition.

Washington Monument, USA

It took 36 years to build the obelisk that stands between the Capitol and the Lincoln Memorial in Washington, DC. Completed in 1884, it forms the keynote of a plan proposed in 1791 by a French officer, Pierre L'Enfant. It was not until 1901 that Congress agreed to implement his scheme. The monument, 555 feet high, can be ascended by elevator.

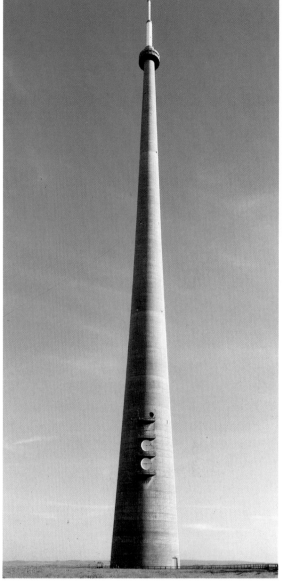

Marine Tower, Yokahama, Japan
The tallest lighthouse in the world stands in Yamashita Park, Yokohama, and measures 348 feet in height. Built of steel, it has a luminous intensity of 600,000 candles and a visibility of 20 miles.

Emley Moor, Yorkshire, England
Built by the Independent Broadcasting Authority, this transmitter is the tallest self-supporting structure in Great Britain, at 1,080 feet. Completed in 1971, it replaced a taller, stayed mast of 1,265 feet which was brought down by icing in March 1969. The mast encases a room at 865 feet and weighs, with the foundations, 14,760 tons.

A Private Xanadu

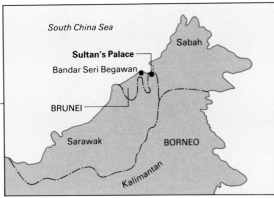

Fact file

The world's largest occupied palace

Architect: Leandro V. Locsin

Built: 1982–86

Materials: Concrete, steel and marble

Floor area: 50 acres

Sultan Hassanal Bolkiak was born on 15 July 1946. He was not expected to become the 29th Sultan, since Hassanal's father inherited the title from his brother in 1953. The throne room (opposite) seats thousands.

The richest man in the world also possesses the longest name. His Majesty Paduka Seri Baginda Sultan and Yang Di-Pertuan, Sultan Hassanal Bolkiah Mu'izzaddin Waddaulah Ibni Al-Markhum Sultan Haji Oamr Ali Saifuddien Sa'adul Khairi Waddien is quite a handle, even if another four lines of honours and titles are omitted . . . Collar of the Supreme Order of the Chrysanthemum, Grand Order of the Mugunghwa, and similar decorations.

The Sultan of Brunei shuns the press, and as a result the details of his magnificent palace, the biggest occupied residence in the world, have largely been obscured in a flurry of lawyers' letters. The least awestruck account of the Sultan and his extraordinary lifestyle comes from the writer James Bartholomew, in his book *The Richest Man in the World*, from which most of the present account is derived.

The Sultan's wealth is based on oil. His small nation of 230,000 people virtually floats on it. It provides free education, health care and subsidized housing for the people, and tops up the Sultan's already bulging bank accounts. According to most estimates, his wealth now amounts to at least $25,000 million, which makes him worth more than General Motors, or ICI, Jaguar and the National Westminster Bank combined. He earns at least $2,000 million a year, or $4.5 million a day—$4,000 dollars a minute. He never has to worry where the next billion is coming from. The estimated $350 million that the palace cost came, in any case, from state funds, not the Sultan's personal fortune.

The Sultan possesses houses all around the world. When in Britain, he often stays at the Dorchester Hotel, which he owns, although he also has a house in Kensington Palace Gardens, another in Hampstead, and a huge property in the suburb of Southall. Once he bought a house near Guildford in Surrey, sight unseen, and drove off to visit it, following another car driven by somebody who knew the way. The two cars became separated, so the Sultan no longer had any idea which way to go. He persevered, however, and in due course reached Guildford. But despite driving around for two and a half hours in search of the house, he failed to find it. He concluded that a house that was so difficult to find was not worth having, so he sold it.

All these houses pale into insignificance compared with his palace in Brunei, the Istana Nurul Iman. The Sultan only decided to build it in the early 1980s, and resolved that it be finished in time for Brunei's independence from Britain early in 1984. Its design and construction were therefore something of a race against time. The architect, Leandro V. Locsin, a distinguished Filipino with uncompromisingly modern tastes, was given just two weeks to come up with a design. The contractors were allowed two years to complete the building, which contains 1,778 rooms. Hardly surprisingly, many things went wrong.

The man who won the contract to build the palace was Enrique Zobel, a Filipino businessman who had met the Sultan playing polo. Zobel persuaded the Sultan that there was no time to put the palace out to tender, and it would be better if he handled the entire job himself. Zobel appointed Locsin, who produced two alternative designs in a hurry. He had never seen the site, or talked to the Sultan, which made his task harder. One design was ultra-modern, the other contained some Islamic motifs and was far less radical. Locsin preferred the first, but the Sultan chose the second. As the design progressed, however, Locsin reverted more and more to his own taste, and away from that of the Sultan.

To help with construction the American project engineers Bechtel were signed up. They recommended that the roof of the throne room, originally to be made of prestressed concrete, should instead be of steel. The demands on this roof are considerable, since as well as spanning a very large area it supports 12 huge chandeliers each of which weighs a ton. There are four thrones, the extra two designed to accommodate a visiting royal couple. Behind the four thrones is a 60-foot Islamic arch, with two further arches inside it, all covered in 22-carat gold tiles.

The banqueting hall, by far the largest in the world, can seat 4,000 people. It, too, has chandeliers and arches tiled in gold. There are 18 lifts, 44 staircases, a total floor area of just over

A Private Xanadu

The palace (above) is situated in a huge compound, seen here beyond the gold-topped Omar Ali Saifuddin Mosque and city of Bandar Seri Begawan. Many of the inhabitants live in kampongs, *water villages in which the houses are built on stilts, like those beyond the mosque.*

Locsin's roof design (above) echoes the roofs of the longhouses that are common in parts of south-east Asia. Much of the effect is lost by not being able to look *down on the roofs, since the palace is on a hill. Locsin was not to know this, having never visited the site.*

50 acres, and an underground car park big enough for 800 cars. The Sultan likes cars, owns at least 110 himself and tends to buy them in threes, so as to be able to vary the colour. He owns several dozen Rolls-Royces, some of which do not have number plates and have never been used. A full-time Rolls-Royce engineer is employed to look after them.

Royal suites inside the palace are provided for members of the royal family, each the equivalent in accommodation to a large house. In all, the suites occupy 900 rooms. Originally, the whole of the interior of the palace was to be covered in marble, but as costs rose that idea was modified. The estimate for the marble alone, ordered by Bechtel, came to $17 million. Alarmed at the size of the bill, Zobel took over the task of buying the marble, altered the specification slightly, and got the bill down to a mere $10 million.

Because the whole building was erected in

The side of the palace overlooking the river (above) has been likened to a multi-storey car park. The palace has 257 toilets and a sewage-treatment plant capable of handling 300,000 gallons a day, adequate for a village of 1,500 people.

such haste, neither architect, contractor nor interior designer ever really discovered what the Sultan's tastes were. Both Locsin and the interior designer, Dale Keller, favoured austere modern designs, but the Sultan's taste runs more to antique or reproduction furniture and decoration of the fancier kind—what designers call "Louis Farouk" style (Louis-Quatorze out of King Farouk). The two styles clashed, and the Sultan has had several rooms completely gutted and rebuilt closer to his own taste.

The entrance to the palace is up a long drive that circles around the entire building before reaching the front, beside which water cascades over huge granite steps. Inside the carved wooden doors, 16 feet high, permanently guarded by two soldiers, is a promenade that runs alongside water. In the middle of the water is an island on which a small orchestra performs. Access is provided by an underground passage

Independence celebrations in 1984 (left) provided the deadline for the palace's completion; in fact, work was still continuing in 1986. The Islamic arches in the principal rooms are covered in gold. The palace is lit by 51,490 light bulbs and 564 chandeliers.

leading to stairs that emerge in the centre of the island. At the end of the promenade, stairs and escalators lead to the main public rooms, the throne room and the banqueting hall.

Architectural critics who were invited to see the palace on its opening were not altogether impressed. *Le Monde* suggested that its pomp was not always in the best of taste. The American magazine *House and Garden* remarked that the rooms did indeed possess astounding glamour and exoticism, but that it was impossible not to be reminded of *The Wizard of Oz* when one looked at the door to the throne room, or of *The King and I* "as one slowly advances toward the gilded, wing-eaved canopy beneath which the Sultan and his Queen sit on state occasions". The most favourable account, inspired by the architect, appeared in the American magazine *Connoisseur*, which accounted the building "a grand success". Yet this irritated the

Sultan, who issued a writ claiming that Locsin was in breach of his contract in talking about the palace to anybody.

It was perhaps as a result of these setbacks that the Sultan decided to build another palace in Brunei, for the use of his second wife, Queen Mariam, a former air hostess. This palace, the Istana Nurulizza, is a smaller, more intimate building, but it still cost $60 million to build and another $60 million to decorate. It includes a high-tech study for the Sultan, with a filing cabinet that descends from the ceiling at the touch of a button, automatically displaying the precise file the ruler has demanded. The couple's son, Prince Azim, had his own personal suite despite being only five years old at the time. It was as pretty as a fairytale, full of knights, castles, and little cottages lost in the forest. But in due course all this elaborate decoration was torn out to be replaced with something more adult.

The reception hall has a marble floor with bold geometric patterns in black. Italian marble of 38 different kinds covers a total of 14 acres inside the palace, while the outside is clad in travertine marble. The Privy Council Chamber is lined with Moroccan onyx, said to be from the very last block of this marble in the world.

The Biggest Church in the World

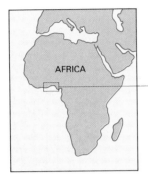

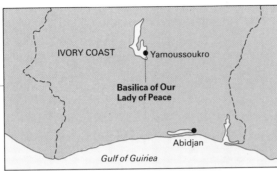

IVORY COAST Yamoussoukro

Basilica of Our Lady of Peace

Abidjan

Gulf of Guinea

AFRICA

Fact file

The world's largest church, modelled on St Peter's, Rome

Creator: President Felix Houphouet-Boigny

Built: 1987–89

Materials: Marble, steel, concrete and glass

Height: 520 feet

Length: 632 feet

Seats: 7,000

The great cathedrals of Europe took whole lifetimes to build, generation after generation of medieval craftsmen working on buildings which they would never see completed. We live in a more impatient age. The biggest church in the world has recently been finished in just three years. What is more, it is not in Europe, close to the sources of marble, steel, concrete and stained glass from which it was built, but in the empty savannah of the Ivory Coast, miles from anywhere. The Basilica of Our Lady of Peace is the grandest architectural gesture of the century, a declaration of faith that cost at least £100 million and that will stand as a beacon to the Christians of Africa—or, as others see it, as the ultimate folly of an old man with intimations of mortality.

The creator of Our Lady of Peace is President Felix Houphouet-Boigny, who as a child of ten had to travel miles to his baptism from his home village of Yamoussoukro, which had no Catholic church. Eighty years later, the basilica in Yamoussoukro is his response, a church modelled on the pattern of St Peter's in Rome. Tactfully, the basilica's dome is a fraction lower than St Peter's, but the crown and golden cross

on top soar to 520 feet, which make it 70 feet higher than the original. It is 632 feet long (20 feet more than St Peter's) while its dome is three times as wide as the dome of St Paul's in London, and the whole of Notre-Dame in Paris would fit inside it several times over. The bronze canopy over the altar is as high as a nine-storey building.

Inside, the basilica has seating for 7,000 and standing room for another 11,000. Outside, the 7-acre marble slab on which it stands provides room for 320,000 more people to participate in services, although the day when that happens is likely to be a distant one. Yamoussoukro, a small town developed by Houphouet-Boigny as the Brasilia of the Ivory Coast, has no more than 30,000 inhabitants, and only about 4,000 of them are Roman Catholics. Abidjan, the capital around which most of the Ivory Coast's population lives, lies 160 miles away to the south—and it already has a modern Catholic cathedral. Even on important feast days, there seems to be little prospect of overcrowding at Our Lady of Peace.

The idea of the basilica occurred to Houphouet-Boigny in 1987. According to the senior engineer responsible for much of the work in Yamoussoukro, Pierre Cabrelli, the President decided that what he wanted to build next was a magnificent church. "I was amazed," Cabrelli told *The Times*. "Who today builds a basilica? Then I asked what was the deadline. The President said that the Pope comes to Africa every four years, that he had been here the year before, and so 'how much time do you have? You work it out'."

The basilica could not have been completed in the time available without modern methods of construction, or without the labour of up to 2,000 men who worked two 10-hour shifts a day. The engineers responsible came from abroad, but the workmen were locally recruited and developed an intense pride in what they had achieved. To criticisms that the basilica was a folly that a poor country such as the Ivory Coast could ill afford, one craftsman retorted: "When they built St Peter's, were there no hungry people in Rome? When England after the Great Fire built itself St Paul's, were there no poor and homeless in London?"

One of the great glories of the building is the dome; the lining of light blue stucco tiles is pierced with 29 million holes that act as a baffle for sound. At its very top are brilliant rings of light blue and dark blue stained glass, leading the eye to the very centre of the cupola, where it finds

Yamoussoukro's local inhabitants are the only people ever to have witnessed construction of so huge a basilica in so short a time; it was erected in just 3 years whereas St Paul's in London took 35 years and St Peter's in Rome 109. Only about 15 percent of the Ivory Coast's population is Roman Catholic; most cling to traditional animist cults.

a representation of a white dove of peace.

The basilica is not the only big building in Yamoussoukro. Houphouet-Boigny, or "Houph" as he is widely known, has transformed his sleepy home town with a vast presidential palace, a marble-fronted conference hall (which has so far housed a single conference), a five-star hotel with its own golf course, three universities and a hospital. The dusty road that leads from Abidjan has been converted into a six-lane highway as it approaches these splendours, most of which stand half-empty amid the encroaching African jungle. While the national university in Abidjan is seriously overcrowded, those in Yamoussoukro, designed in *grande-école* French style, are magnificently empty.

Classical formal gardens like those at Versailles have been created as a setting for the basilica, replacing contractors' temporary buildings *(above)*. The area of stained glass *(left)*— 80,000 square feet— eclipses any other ecclesiastical building. Over 4,000 shades of glass produce dramatic effects in the 10-storey high windows.

The Biggest Church in the World

The 272 columns, using Corinthian, Doric and Ionic orders, reach up to 100 feet and were built of concrete to save time and money. Other elements of the building also use modern methods—the dome is of light grey anodized aluminium over polyurethane insulation. Modern techniques are employed to create a Renaissance structure.

Equipped with everything the student could desire, they teach only engineering and agriculture.

In many other countries, such fantastic spending would condemn Houphouet-Boigny as a megalomaniac intent on destroying his country in order to create a memorial to himself. But many of those who come to Yamoussoukro to mock find the criticisms dying in their throats. It is true that, according to figures from UNICEF, the money spent on the basilica might have been used to vaccinate the Ivory Coast's 10 million people against six diseases—diphtheria, measles, whooping cough, polio, tetanus and tuberculosis—which claim thousands of victims every year. Furthermore, the Ivory Coast faces economic catastrophe, with a debt of $8 billion on which payments have had to be suspended. But for some Houphouet-Boigny's own sincerity and the splendour of what has been achieved make such criticisms seem irrelevant.

Until 1980, Houphouet-Boigny had achieved remarkable things for his country, once a colony of France. While many newly independent nations in Africa had degenerated into tribal conflict and poverty, the Ivory Coast was a huge success. An economy based on cocoa, coffee and cotton had created wealth and stability. The regime, though dominated by a single man, was benign and liberal. Then the prices of raw materials began to decline, as the world turned away from chocolate. To pay the guaranteed prices to the cocoa growers, the Ivory Coast ran up the highest per capita debt in Africa. Instead of being able to present Yamoussoukro as the crowning glory of his success, Houphouet-Boigny was forced to defend it against harsh criticism, both internal and external.

He claimed that all the money it had cost had come from his own purse, a suggestion that provokes some chuckles but is not wholly impossible. He was already a rich man when he became president, and is said to have invested the family wealth skilfully. Taxed with the suspension of debt repayments in 1988, when the cost of the basilica was put at £80 million, he asked "How could my little £80 million help?" When criticisms came from France, he was exasperated: "How can a people who are proud of Versailles, of Notre-Dame, of Chartres, not understand?" he pleaded. The Ivorian Minister of Information, M. Laurent Dona Fologo, was less subtle. He called the criticism "racist" because, he said, critics obviously "cannot stand

to see Africans with something big, beautiful, and lasting".

In part to deflect criticism, in part to ensure it will survive him, Houphouet-Boigny offered the completed basilica to the Vatican, a gift which threw the civil servants of St Peter's into something of a tizzy. How could they turn down such a magnificent expression of the Catholic faith in a continent full of unbelievers? After three months' thought, and some negotiation, they accepted the gift but only on condition that the upkeep will be borne by the Ivory Coast. A special fund has been established by Houphouet-Boigny in the Vatican Bank from which will be drawn the £1.7 million a year the upkeep of the basilica is expected to cost. The Vatican is also believed to have extracted a promise from the President that he will increase spending on his people's health and education. In return,

Precast concrete was the main material used to build the basilica, the sections being erected by 6 cranes running on rails (left). Granite was imported from Spain, steel from Belgium, marble for the peristyle from Italy and glass from France.

The basilica's rotunda with the cupola ready to be placed on top of the dome. Although the dome itself is slightly lower than St Peter's, the cupola and golden cross make the overall height greater. At night the dome is lit by 1,810 1,000-watt lights.

Houphouet-Boigny persuaded the Pope to visit Yamoussoukro in January 1990 to inaugurate the basilica.

Whether the building is seen as a gigantic caprice or as a beacon for Africans depends on the eye of faith, and on what becomes of it after Houphouet-Boigny dies. When the cathedral was finished at the end of 1989 he was at least 84 (some say over 90) and had been president for 35 years. He was unwilling to discuss the question of his successor until the basilica was completed for fear that he might not survive to see it finished were he to reveal his intentions. Once his influence is gone, can the unkind climate of Africa be kept at bay? Will the faithless poor be prepared to sustain a dream in which they have no share? The history of Our Lady of Peace is likely to be just as interesting and unpredictable as the story of its construction.

A window frame about to be lifted into position above the portico. The air-conditioned basilica requires a maintenance staff of 25, of whom 8 do nothing but wash and polish the marble of the square and peristyle on which an open-air congregation would stand. Others spend all day polishing the 7,000 reddish-brown iroko hardwood pews.

World within a World

Fact file

The first large-scale attempt to replicate Earth's behaviour and environment

Architect: Margaret Augustine

Built: 1987–90

Materials: Steel and glass

Area: 3.1 acres

God created the Earth in six days, according to the Book of Genesis. In the desert 40 miles north of Tucson, Arizona, a team of scientists and visionaries is attempting to reproduce that feat. They are taking a little longer and may not get quite such a satisfactory result; but the project is testing techniques of building to the limit and if it works will provide a mass of information that may be useful in managing life on Earth, and in planning new settlements in space.

The project is called Biosphere II, and the object is to produce a completely sealed environment which will simulate the behaviour of the Earth. Within two huge glasshouses, an entire ecosystem will be created, providing all the food, water and air needed to keep alive eight people who will spend up to two years inside. Air, water and wastes will all be recycled, and natural weather patterns mimicked. Nothing will be allowed to enter or leave the sealed container, which consists of two linked space-frame structures covered in glass.

Inside one will be a five-storey building in which the experimenters will live, a tropical rainforest, a region of desert, an ocean complete with a coral reef and tides, a savannah region, and a selection of birds, reptiles, insects and small mammals. The other structure will be used for intensive agriculture. In all, some 3,800 plant and animal species will inhabit Biosphere II, whose name reflects the fact that it is not the first experiment of this kind. Biosphere I is the Earth, that abused but still functioning planet the Arizona pioneers are seeking to mimic.

Building such a revolutionary structure has created considerable problems. One of the most difficult is ensuring that the inside is completely isolated from the outside. A normal well-sealed building exchanges its air completely one to three times every day. Biosphere II is designed to exchange all the air within it only once in its entire lifetime, which is set at 100 years. "No matter how you define airtight, that is still an awesome specification," says Peter Pearce, president of Pearce Structures Inc, a California firm that won the contract to design and build the space frame. It is the first time that any builder has ever attempted to achieve such a result. Pearce says that it is more like building a space shuttle than a building.

On top of the concrete foundations a thin stainless-steel liner has been inserted, rising at the edges to form a seal to the glass walls of the building. The basic structure consists of ribs of steel, linked together in a novel way and with the panes of glass sealed directly to the steel with a silicone material. The design was produced by Pearce in four months, after scientists had realized that no existing glazing system was capable of doing the job. The structural principle that lies behind it is the space frame, a very strong and flexible building system based on attaching steel tubes together in triangular shapes. Space frames are inherently stronger than traditional structures, use less material, and can be built in many different shapes. One of the two structures in Biosphere II—the agricultural space—consists of a series of circular vaults, while the other is pyramidal in form.

The organization behind the project is called Space Biosphere Ventures, and it has been funded by Edward Bass, a Texan who has invested $30 million to get Biosphere II built. The architect of the building is Margaret Augustine, a British architect who is also Chief Executive Officer of SBV. When finished, Biosphere II will cover an area of 3.1 acres, rise to 85 feet at its highest point, and enclose a volume of 7 million cubic feet of air. In addition to the apartments for the eight researchers, it will incorporate laboratories, computer and communications facilities, workshops, libraries and recreation

facilities. The "biospherians", as they call themselves, will not be cut off totally from the outside world; they will be able to watch TV, listen to the radio, or talk to colleagues outside by telephone.

The eight biospherians are hoping to enter the sealed environment in September 1990, if the builders have finished it by then. Once inside, they will breathe oxygen created by the plants around them. Water will evaporate and form "clouds" inside the glass, rising to the highest point in the building where cooling coils will condense it back into water. Then it will flow down the 50-foot high mountain an through a miniature rainforest into an artificial ocean, which is 35 feet deep, and from which evaporation will occur. The only thing that will come from outside is energy, and information.

The Sun will supply much of the energy, but electricity will also be fed in to run various mechanical systems, including the artificial wave-maker which is necessary if coral is to grow in the ocean. The heat of the Sun will be controlled by louvres that can open and close. Botanists are hoping that the steel parts of the space frame, which cut out some light, will not prevent plants from growing normally.

Most of the food eaten by the inhabitants of Biosphere II will come from intensive farming techniques developed at the University of Arizona. A total of 140 different crops will be grown, with waste products recycled to provide nutrients. Vegetables, including cucumbers, tomatoes, lettuce and broccoli, fruits such as

The unique outlines and space frames of Biosphere II were influenced by Buckminster Fuller, whose work is admired by the designer of the frame, Peter Pearce. The various "biomes" will replicate rainforest, tropical savannah, salt- and fresh-water marsh, desert, ocean and a thorn-scrub forest, as well as intensive agriculture.

World within a World

The aquaculture bay (below) provides the surroundings for lunch. The "biospherians" will eat some of the fruit and fish that the biosphere will provide. Milk will be produced by African pygmy goats and chickens will supply eggs and meat.

Hydroponic tomatoes: the root systems are kept in darkness and draw their nourishment from water impregnated with nutrients. About 20,000 square feet will be devoted to intensive agriculture. Produce from the greenhouse is sold to cover costs.

papaya, bananas and strawberries, and cereal crops such as wheat, barley and rice can all be grown in this way. Fertilizer will be provided from the waste products of the fish grown in the fish farm. The builders have been obliged to avoid using any curing agents in the concrete foundations, for fear that it would migrate into the soil and poison the occupants, or at least interfere with their experiments.

Biosphere II is not the first attempt at survival in a totally enclosed system. Inhabitants of spacecraft face similar problems, and experiments have been carried out on the ground in the Soviet Union since the 1960s. Large bottles containing sealed gardens are a simple example of the principle, and one of those, at the University of Hawaii, has been functioning

successfully without human intervention for 20 years. SBV have done their own experiments, in a much smaller test module designed to test the principles.

In March 1989 a marine biologist, Abigail Alling, emerged unharmed after five days inside the module, which SBV claims is the longest time any human being has ever occupied a completely sealed ecological system. The object of the experiment was to test the systems for controlling the accumulation of toxic gases. Some of the Soviet experiments had to be terminated when toxic gases such as sulphur dioxide, nitrogen dioxide, ammonia, carbon monoxide, ozone and hydrogen sulphide increased to a dangerous level. In Biosphere II, such gases will be controlled by forcing the air through soil con-

The aquaculture bay will generate some food for the biospherians, but it will also provide nutrients for crops in soil beds irrigated by fish water. These nutrients result from the excretion of ammonia in fish wastes; the naturally present bacteria in the biofilter tank convert this to nitrates which are used to fertilize crops.

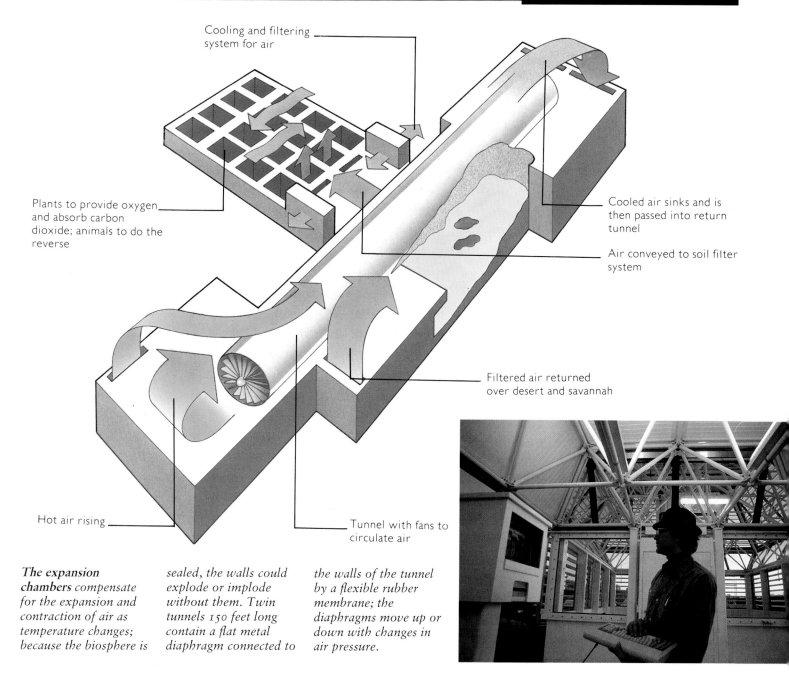

Cooling and filtering system for air

Plants to provide oxygen and absorb carbon dioxide; animals to do the reverse

Cooled air sinks and is then passed into return tunnel

Air conveyed to soil filter system

Filtered air returned over desert and savannah

Hot air rising

Tunnel with fans to circulate air

The expansion chambers compensate for the expansion and contraction of air as temperature changes; because the biosphere is sealed, the walls could explode or implode without them. Twin tunnels 150 feet long contain a flat metal diaphragm connected to the walls of the tunnel by a flexible rubber membrane; the diaphragms move up or down with changes in air pressure.

taining microbes which should convert the toxic gases into harmless chemicals.

The air inside the enclosure will be circulated mechanically, since it is impossible to simulate the movements of the atmosphere on such a small scale. A set of 3,500 sensors have been installed to measure the conditions and to control the rate of air circulation. Monitors have also been installed to ensure that the building really is as leak tight as it is supposed to be.

The experiment has at least two purposes. One is to try to understand better how the Earth itself functions, what are the critical parameters, and how it might be better managed. A second is to lay the ground rules for constructing habitable colonies in space. If such colonies are to function successfully, they will need to be self-supporting,

since transporting food and fuel from Earth will be far too expensive. That would be like expecting the colonists who settled in the New World in the seventeenth century to be supplied for ever by convoys of vessels crossing the Atlantic from Europe.

The whole concept of Biosphere II is futuristic. And since it is not the kind of thing many official research organizations would dare to finance it needed the patronage of a philanthropist to get it started. But many respectable scientists have been consulted, and take the idea seriously. One of them, Carl Hodges, Director of the Environmental Research Laboratory at the University of Arizona, admits it may not work. "We could get the balance wrong," he says. "It's a big step, but big steps are the most fun."

A test module for the biosphere, of 17,000 cubic feet, was built to test materials, techniques and the computer monitoring systems. By using it to test the closed-system dynamics of ecosystems that will be replicated in the biosphere, many potential problems were overcome. Each series of experiments was for a 3-month period.

CHAPTER FOUR

Feats of Civil Engineering

Human mastery over nature is measured by the achievements of civil engineers, the men (and increasingly women as well) who bridge rivers, dam lakes, build roads and railways, dig canals and erect coastal defences to defeat the power of the sea. Once finished their works are easily taken for granted, unless they should fail, when the air is loud with lamentation. For unlike other forms of engineering the work of civil engineers is expected to last for ever, a permanent modification of the natural world. They must work for posterity as well as for their clients.

Some civil engineering works are so huge and permanent that they even outlast the purposes that gave rise to them. The Great Wall of China, perhaps the most remarkable construction in the history of civilization, still leaps from crag to crag across the immensity of that country, although the threat it was meant to counter has long disappeared. The Panama Canal altered geography permanently, even if ships should one day cease to use it. A desert in the state of Washington, arid and uncultivable since man first discovered it, was turned into productive farmland by the Grand Coulee Dam. And the coastline of the Netherlands has been remodelled radically—and, the Dutch hope, permanently—by the Delta Plan, one of the greatest but least heralded feats of civil engineering this century.

Civil engineering can also change perceptions. The nation of Canada might not have survived at all but for the long ribbon of steel, in the form of the Canadian Pacific Railway, that stitched it together. The dream of capitalizing on the riches of Siberia would have remained even more remote without the Trans-Siberian and BAM railways. And the concept of a united Europe would seem less realistic but for the railways and roads that tunnel beneath the barrier of the Alps, linking Germany and Switzerland to Italy.

As the world's reserves of fossil fuels diminish, the search for alternative sources of energy is bound to accelerate, with schemes like the Orkney turbine to exploit wind power, and the solar oven in France to harness the power of the Sun.

The tasks of civil engineering, for all their permanence, are never complete. As one great project ends, another, even more ambitious, swims into view, made possible by improving techniques. Since the very first roads and bridges were built, humans have pursued a desire to adapt and utilize nature, and there is no sign we will ever desist.

Feats of Civil Engineering
Great Wall of China
Panama Canal
Canals across the world
Canadian Pacific Railway
Trans-Siberian Railway
Grand Coulee Dam
Dams: harnessing the power of water
Dutch Delta Plan
St Gotthard Pass
The world's greatest highways
Ironbridge
Humber Bridge
Bridges of distinction
Statfjord B Oil Platform
Orkney Wind Generator
CHOOZ-B Power Station
Odeillo Solar Oven

The Longest Bastion

Described as the greatest construction enterprise ever undertaken, the Great Wall took 20 centuries to complete and refine

Original builder: Qin Shi Huangdi

Built: 3rd century BC–17th century AD

Materials: Earth, stone, timber and bricks

Length: 2,150 miles

Qin Shi Huangdi (221–210 BC), the first emperor of China, who began the Great Wall. Though his reign was brief, Qin established the political form by which China was to be ruled until 1911. He is best known for the army of terracotta horses and warriors found near his tomb in Xi'an.

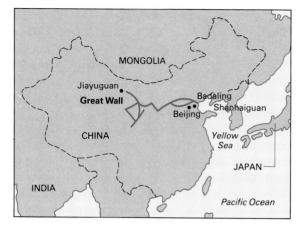

The greatest construction project ever carried out by man stretches 2,150 miles across China along a twisting, curving course that has been compared to the body of a dragon. Created over a period of more than 1,800 years by millions of soldiers and workmen, the Great Wall of China runs from the Yellow Sea near Beijing to the Jade Gate at Jiayuguan, which marked the outer limits of Chinese influence and the beginning of the Central Asian wilderness. The wall formed the boundary between Chinese civilization and the barbarians lying to the north, the point at which the spread of Chinese culture finally petered out in the mountains and deserts where nomads eked out a precarious existence. The Great Wall thus represents "the most colossal tide-mark of the human race" in the words of the American scholar Owen Lattimore.

The wall was begun in the reign of the First Emperor, Qin Shi Huangdi, who waged a war of conquest and finally united China in 221 BC. Before that, as early as the fifth century BC, there had been smaller walls built by local rulers, many of which were destroyed by Qin. He established a ruthless and efficient empire, with a system of criminal justice, a network of new roads, and a bureaucracy that controlled where people lived and how far they could move. Criminals were treated harshly, while those unwilling to work were drafted into the army and despatched to the farther corners of the empire. It was these people who were the first builders of what we now call the Great Wall.

According to contemporary history books, Qin sent his top general, Meng Tian, at the head of an army of 300,000 men to put down the barbarians in the north and to build a wall following the terrain, using natural obstacles and passes to form an impenetrable barrier.

The wall we see today is, however, mostly of much later construction, dating from the Ming dynasty (1368–1644). Its purpose was the same as that of Qin's—to prevent invasions from the north and mark in an unambiguous way the borders of the empire. The best-preserved section of the Ming dynasty wall lies between Beijing and the sea, a 400-mile stretch of masonry running eastward along the high ridges of the Yanshan mountains to Shanhaiguan.

In between these two periods, other rulers of China also set their stamp on the wall, putting millions of unwilling labourers to work on different sections. Qin Shi Huangdi used his army, plus half a million peasants, to create his wall. More than 600 years later, in AD 446, Taiping Zhenjun called up 300,000 labourers to build another section, while in AD 555 Tian Bao press-ganged 1.8 million peasants for work.

There were periods when interest in the wall declined; the Tang dynasty, which began in AD 618, considered attack the best line of defence, and created a strong army rather than reinforce the wall. But when the Mings took over, the wall regained its priority in the scheme of things. The wall we see today is thus the product of millions of men, but a single idea.

The materials used include earth, stone, timber, tiles and, during the Ming dynasty, bricks. Because transportation was difficult, local ones were employed: stones in the high mountains; earth on the plains; sand, pebbles and tamarisk twigs in the Gobi Desert; oak, pine and fir from forests around Liaodong in the north east. Many of these materials made an impermanent wall; thus it is the stone and later brick-and-tile sections that have survived. During the Ming dynasty, kilns were built on the spot to create the bricks and tiles, as well as the lime used to bind them together.

The building materials were carried by human effort, on a man's back or on carrying poles. Sometimes the labourers formed a human chain, passing the stones or bricks from hand to hand up the mountain side. Handcarts were also used, and large rocks were hauled with windlasses or moved with levers. Donkeys carried baskets loaded with bricks and mortar, while goats are also said to have been pressed into service, the bricks tied to their horns.

The earthen walls of the Qin dynasty were made by erecting shuttering in the forms of posts and boards along the line of construction, and then filling the space between with soil. Layers of earth 3–4 inches thick were laid and then

The Longest Bastion

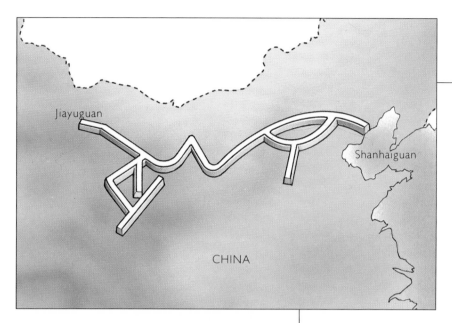

Watchtower

hammered solid with mallets before laying subsequent layers. Similar methods were used in the Ming period, except that each successive layer was deeper—by 8 inches or so. This technique was well developed in China, since it was often used for building the walls of houses.

Masonry sections were built by first levelling the ground and laying a series of courses of stone slabs to form the foundation. The faces of the wall were then constructed from stone, and the gap between them filled with small stones, rubble, lime and earth. Once the wall was tall enough, the top was made in the form of layers of brick, either laid on a slant for gentle inclines or in the form of a staircase if the slope was more than 45 degrees.

One of the most astonishing features of the Great Wall is the way it makes use of the defensive qualities of the ground, often striking out along a ridge, only to curve back on itself to follow the natural features and dominate the high ground. At key points forts and towers were built from which to survey the land commanded by the wall. These points naturally tended to be the places where the enemy would seek to attack—mountain passes, road junctions, or bends in a river on flat land. As an encyclopedia written during the Tang dynasty puts it: "Beacon towers must be built at crucial points of high mountains or at turning points on flat land."

Although the function of the wall was defensive and utilitarian, many of its details were designed with a real sense of style. Towers, gates and forts were often beautifully detailed, in a wide variety of architectural styles. There were

The route of the wall is not a single line, since it incorporates a series of walls built by successive rulers. The early walls were rudimentary and had to be rebuilt.

The wall clings to the ridges as it snakes over the mountains. Wherever the gradient was less than 45 degrees, bricks lining the walkway followed the contour of the wall; where it exceeded that angle, steps were created.

Wall section
Height: 20–30 feet
Width: 25 feet at base
20 feet at top

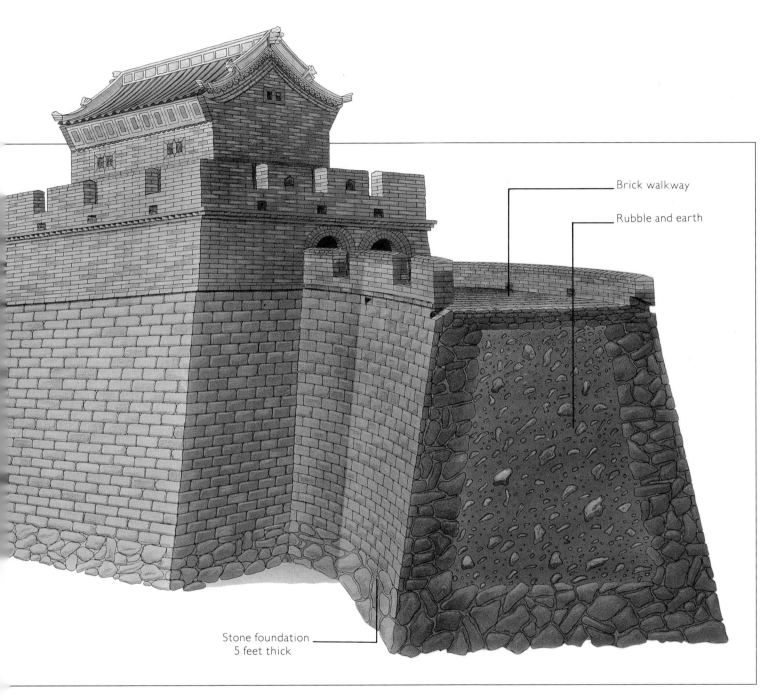

Brick walkway

Rubble and earth

Stone foundation
5 feet thick

Watchtowers

Generally 40 feet square in plan and up to 40 feet high, watchtowers are estimated to number about 25,000. Beacon towers, placed at a maximum distance of 11 miles, used wolves' dung, sulphur and nitrate to produce a column of smoke to indicate to adjacent towers the strength of an attack. At night dry timber was used.

also temple buildings and shrines along the wall, as well as tea-houses and clock towers.

The wall itself stood about 22–26 feet high, and equally broad at its base, narrowing to about 16 feet at the top. (These measurements apply to the best-preserved section of the wall, near Beijing, which dates from the Ming period.) Every 200 yards or so, an archway on the Chinese side of the wall gave access to a staircase leading to the top of the wall; the top served as a pathway as well as a line of defence, enabling troops to move swiftly, ten abreast, along its length to reinforce a garrison under attack.

On the inner edge was a 3-foot parapet to reduce the risk of falling off, and battlements up to 6 feet high provided protection on the outer edge. Every 100–200 yards was a projecting platform, forming a buttress to strengthen the wall, and a place from which soldiers could bring fire to bear on enemies trying to climb the sides of the wall.

At similar distances were placed ramparts, two- to three-storey buildings in which the soldiers lived. About 30–40 feet high and 120–180 feet square, the ramparts were topped by an area from which cannons could be fired. Each rampart was garrisoned by between 30 and 50 soldiers under the command of a petty officer.

When an attack came, the soldiers used a well-established system of signals for indicating its

The Longest Bastion

general in command. When an attack came, all nine were brought under the command of the Minister of War. Each military zone had its headquarters in a city along the wall, or in an important fort, in close communication with the capital. At its peak, the system worked well.

To the Ming dynasty, the wall was all that stood between them and the Mongol hordes, who had begun the creation of their empire under Genghis Khan at the beginning of the thirteenth century. Despite the Mongols' small numbers and modest army of only 250,000 men, their ferocity enabled them to breach the Great Wall and conquer China. By the end of the thirteenth century their empire stretched in a great swathe across Asia and Europe, from Korea to Poland and Hungary in the north, and from south China to Turkey in the south.

The great emperor Kublai Khan, grandson of Genghis, emerged as emperor in 1260, ruling China with considerable skill. After his death, however, Mongol dominance began to crumble, until the Mongols were finally thrown back behind what remained of the wall and Zhu Yuanzhang established the Ming dynasty.

Given this background, it is hardly surprising that particular attention was paid to reinforcing the wall as the only way of preventing another successful assault from the Mongols. Zhu sent his nine sons north to head the nine garrisons defending the wall, and more and more fortifications were built. Building continued throughout the entire Ming dynasty, and the bulk of the wall as we know it today was created during that period, between 1368 and 1644.

A wall so prodigious as that built by the Chinese could not fail to charm Europeans, once they became aware of it. Dr Johnson was especially enthusiastic, and would have liked to have visited the wall himself. One day James Boswell, his biographer, remarked to Johnson that he too would like to visit the wall if he did not have his children to look after. In a famous retort, Johnson remarked: "Sir, by doing so, you would do what would be of importance in raising your children to eminence. There would be a lustre reflected on them from your spirit and curiosity. They would at all times be regarded as the children of a man who had gone to view the Great Wall of China. I am serious, Sir."

In 1909 the American writer William Edgar Geil became one of the first westerners to follow the entire course of the wall, and became immensely enthusiastic. He claimed that the

importance. One beacon fire and one salvo meant between 2 and 100 enemy; two beacon fires and two salvos meant up to 500, and so on, up to five fires and five salvos, which meant an attack by more than 10,000 enemy troops.

In times of peace, the soldiers grew their own food on land close to the wall, so as to remain self-supporting. They kept guard, and monitored the movement of merchants crossing the wall with goods for sale. The upkeep of the wall was their responsibility, and they were given strict instructions about how to maintain it. Their weapons included gunpowder, invented during the Ming dynasty, which was used for various types of grenade. Real artillery was not then available, or it would have proved much more difficult to defend the wall, but the siege crossbow and a version of the Roman catapult were used for firing large projectiles over considerable distances. Swords, spears and cudgels were effective in close combat, and mounted cavalry were also employed.

The length of the wall, claimed during the Ming period as 10,000 li (4,000 miles), was divided up into nine military zones, each with a

The Great Wall was seen by the Chinese as a dragon with its head drinking from the Bo Hai Sea at Shanhaiguan (above). The head is a section 76 feet long which protrudes into the sea, and was damaged by a British force invading in 1900 to quell the Boxer uprising. The gate at the first fort is known as "The First Pass Under Heaven".

The wall at Gubeikou in Hebei Province, where a military unit was set up under the Ming dynasty to guard a dozen passes, of which the most important was Gubeikou itself. It was here that the first British Ambassador to China first saw the wall, in 1793.

The unrestored sections of the wall vary in dilapidation from ridges of earth to stone sections that require little restoration. This example of unrestored wall is adjacent to the restored section visited by tourists at Badaling, east of Beijing.

builders of the wall had been ahead of the senseless militarism of Europe. In the 1980s, more than four million people every year have followed Geil's footsteps, though few have ventured further than the well-preserved section of wall within an easy coach trip from Beijing. One who did was William Lindsay, a university research worker from Merseyside, England, who was hailed by the New China News Agency in 1987 as the first foreigner to run unescorted and alone along 1,500 miles of the Great Wall. The journey, a combination of running, walking and limping, took 78 days, with a four-month break in the middle.

During the twentieth century, the Chinese have lacked the resources to maintain the wall in the manner it merits. Some short stretches are magnificent, but other parts, out of reach of tourists, have been allowed to decay. The wall today no longer has any defensive role, though Chinese troops used it as a highway in the war against the Japanese. But it remains one of the wonders of the world, a work of man on the scale of nature, an astonishing example of human strength, ingenuity and endurance.

Joining the Oceans

Fact file

The world's largest and most costly engineering project when built

Engineers: de Lesseps, John Stevens, George W. Goethals

Built: 1881–89, 1904–14

Length: 51 miles 352 yards

The building of a canal to link the Atlantic and the Pacific was the largest and most expensive engineering project ever undertaken. It took more than 40 years from its conception to the first transit by an ocean-going ship. It employed tens of thousands of men and broke new ground in engineering, planning, medicine and labour relations. It was the final fling of nineteenth-century European optimism—and the first evidence that the United States had become a great power. It changed geography, dividing a continent to unite the oceans and even helped to create a new nation, Panama.

The story began in 1870, when two ships of the US Navy were despatched to the Isthmus of Darien, the narrow neck of land joining the Americas, to establish where a canal might be dug. The objective was clear: from New York to San Francisco around Cape Horn was a journey of 13,000 miles, taking a month to complete. Through a canal it would be only 5,000 miles. Yet the difficulties of cutting a canal through this narrow strip of land—only 30 miles across at its narrowest point—could hardly be overestimated. Driving a railway across it in the 1850s had taken five years and cost six times the estimate. Thousands had died of cholera, dysentery, yellow fever and smallpox.

However, before further action was taken by the US, a group of French financiers obtained a concession to construct a canal from Colon to Panama. Their choice of engineer was Ferdinand de Lesseps, a French diplomat and politician who had made his name building the Suez Canal. He arrived in Panama City in 1880 and after a cursory survey decided to build a canal along the Chagres River and the Rio Grande, linking the Atlantic and Pacific at sea level and following closely the line of the railway. No sooner had work begun than people began to die. Panama was a hell-hole, one of the unhealthiest places on earth: mosquitoes bred in millions in swamps and pools of water; proper drainage did not exist; and the medical knowledge of the French

pioneers was simply inadequate. One of those who laboured on the canal was the French painter Paul Gauguin, who arrived in 1887 full of dreams of buying land and living for nothing on fruit and fish. He hated the place, and the people. As soon as he had made enough money, Gauguin left for Martinique.

In 1889 de Lesseps' company crashed, his dream of a canal at sea level having proved impracticable. Failure was followed by scandal: Ferdinand de Lesseps was accused of corruption, and the collapse of the company became a French *affaire* in which governments fell and reputations were ruined. A total of $287 million had been spent—far more, at the time, than had ever been spent on any peacetime operation—at least 20,000 had died, and just 19 miles of canal had been dug. It was a deep and humiliating failure for France.

By the turn of the century, however, the United States had regained enthusiasm for the canal, and opened negotiations with the Government of Colombia, of which Panama was then a department. The Colombians rejected a proposed treaty but, with tacit encouragement from President Theodore Roosevelt, a group of Panamanians declared themselves independent. Within two days, their government was recognized by Washington. For the sum of $10 million, plus $250,000 a year from 1913 onwards, Roosevelt won agreement from the new government to dig the canal.

The first task, without which all else would fail, was to control disease in Panama. An Army doctor, Colonel William Gorgas, was appointed to take charge of the hospitals and sanitary arrangements. It was an inspired appointment, for Gorgas had already been largely responsible for eliminating yellow fever in Cuba, by controlling the mosquito *Stegomyia fasciata*. By eliminating both this and the *Anopheles* mosquito, responsible for malaria, Gorgas transformed the prospects for success in Panama.

But there was still a canal to be dug. Unlike de

Gatún Lake, bisected by the canal channel, was created by a dam and sets of locks at each end and forms the central part of the canal. The most formidable challenge to the builders was Culebra Cut (right), now called the Gaillard Cut. Every day 60 steam shovels filled trains with spoil; often they were engulfed by the collapse of the channel walls, which became a more frequent occurrence from 1911 as the cut grew deeper.

Lesseps, the American engineers made no attempt to build a canal at sea level. Instead, they designed a series of locks to carry the ships up and down, a far more practical scheme. But the amount of digging to be done was still prodigious. The toughest section of all was the Culebra (now Gaillard) Cut, an 8-mile stretch between Bas Obispo and Pedro Miguel. Here a man-made canyon was created through a mountain, driven by steam shovels, explosives and a workforce of 6,000 men. It took seven years and 61 million pounds of dynamite, more explosive energy than had been expended in all the wars the US had ever fought. The noise was tremendous, the dangers enormous, the loss of life barely calculable.

The trouble was the instability of the rock. As it was removed from the sides of the cut, the

Joining the Oceans

Ships pass from the Caribbean Sea to the Pacific Ocean in an easterly direction.

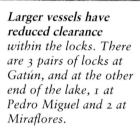

Larger vessels have reduced clearance within the locks. There are 3 pairs of locks at Gatún, and at the other end of the lake, 1 at Pedro Miguel and 2 at Miraflores.

Electric locomotives at each lock act as land tugs, hauling ships through, using a powered windlass and 800 feet of steel cable, with the smoothness of a piston through a cylinder.

A cross-section through the canal illustrates the rise of the canal to Lake Gatún, whose surface level ranges from 82 to 87 feet above sea level. The surface of Miraflores Lake is at 54 feet above sea level.

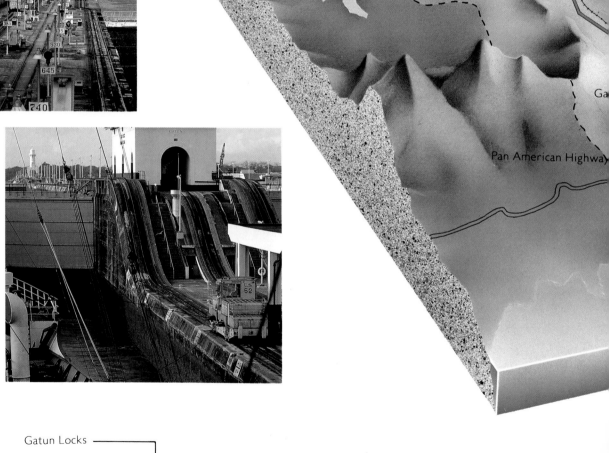

Gatun Locks

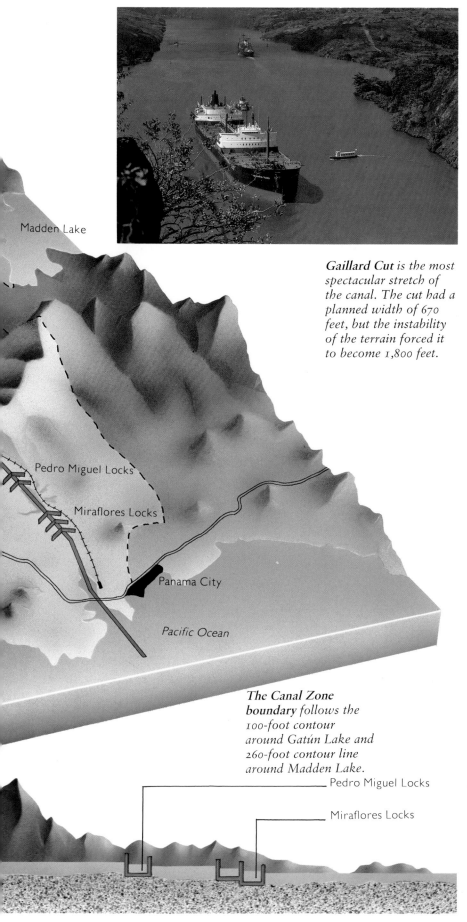

Gaillard Cut *is the most spectacular stretch of the canal. The cut had a planned width of 670 feet, but the instability of the terrain forced it to become 1,800 feet.*

Madden Lake

Pedro Miguel Locks

Miraflores Locks

Panama City

Pacific Ocean

The Canal Zone boundary *follows the 100-foot contour around Gatún Lake and 260-foot contour line around Madden Lake.*

Pedro Miguel Locks

Miraflores Locks

walls literally began to bulge as pressure drove them outward. The same effect even caused the floor of the cut to rise, sometimes astonishingly fast, by 15 to 20 feet. It was disheartening indeed. Everything was tried to stop the falls, including plastering the walls with concrete, but all failed. The concrete crumbled and fell, together with the rock. The only way was to reduce the slope of the walls until stability was achieved, and that meant the intended narrow defile becoming a saucer-shaped depression.

The great chain of locks at each end of the canal, the biggest such structures ever built, were another source of wonder. Stood on end, these locks would tower above most buildings in Manhattan today, except for the Empire State, the World Trade Center and a few others. Yet they are not simply buildings, but machines which work as smoothly as a sewing-machine. It took four years to build them, beginning in August 1909. They were built in pairs, to take two lines of traffic at once.

The locks are made of concrete, poured into huge wooden forms. The floor of each chamber was 13 to 20 feet thick, while the walls were as much as 50 feet thick at floor level, narrowing in a series of steps on the outside to only 8 feet at the top. The total quantity of concrete poured to make all 12 chambers was 4.4 million cubic yards. The concrete remains in perfect condition.

The walls of each chamber are not solid, but threaded with huge passageways through which water runs to fill and empty the chambers. The water comes from Gatún Lake and Miraflores Lake and is fed into the bottom of each set of chambers, where it flows through 70 holes to lift the ships smoothly and gently. A similar set of holes is used to drain the chambers, lowering ships if they are moving in the other direction. The flow of water is controlled by sliding steel gates, running on roller bearings, which can be raised and lowered across the culverts.

At the end of each chamber are huge gates, each weighing hundreds of tons. Made of steel plate riveted to a backbone of steel girders, they swing together to form a flattened "V". They were designed to float, so that in use they would exert the least possible force on their hinges. Each leaf is 65 feet wide and 7 feet thick, but their height varies with their position. The largest, at Miraflores, are 82 feet high and weigh 745 tons.

The life-blood of any canal with locks is water. Because Panama is a country with ample rainfall, until recently there seemed no danger

Joining the Oceans

that the canal would run dry. However, the lakes rely on the surrounding rain forest for replenishment of the water lost to the sea each time a ship passes through the canal. Extensive felling of the forest has reduced the flow of water into the lakes by destroying the rain forest's ability to act as a giant sponge; there is now serious concern that a shortage of water will jeopardize the canal's future operation.

The flow of water over the spillway at Gatún is used to generate electricity, which in turn powers everything else on the canal—the valves, the lock gates, and the small, specially designed locomotives which travel along tracks laid at the top of the lock chambers, towing the ships through. No ship was to travel through the locks under its own power, for fear that it would get out of control and smash through the safety chains and into the lock gates.

Everything, except the operation of the locomotives, was designed to be under the control of one man and a single control panel. On this panel are depicted all the functions of the locks—the gates, the valves, the water levels, and next to each is a simple control switch. Switches can only be moved in the correct order, so that it is impossible, for instance, to try to open the lock gates against a head of water.

The control panels have ensured that the canal has worked like clockwork since the day the first ship went through it. That took place on 7 January 1914, with little ceremony, when an old French crane boat, the *Alexandre la Valley*, made the first complete transit through the canal. On 3 August the *Christobal*, a cement boat, became the first ocean-going ship to travel from ocean to ocean, and the *Ancon* was the first passenger ship to cross, on 15 August. But the Culebra Cut still had problems to present, and in October a huge collapse blocked the entire canal. There were further falls in 1915, and dredging has continued to this day.

Ten years after it opened, more than 5,000 ships a year were passing through the canal, and by 1939 the number reached 7,000. After World War II it doubled, and hit a peak of about 15,000 ships a year in the early 1970s. The greatest toll ever paid by a ship was $42,077.88 by the *Queen Elizabeth II* in March 1975; the smallest, 36 cents, by Richard Halliburton, who swam through the canal one day at a time in the 1920s. He even persuaded the authorities to let him swim through the locks, and, like any other vessel, paid a toll based on his weight.

Creators of the Panama Canal

An enterprise on the scale of the Panama Canal requires men of rare calibre to see it through to success. Four such men became involved with its construction.

President Theodore Roosevelt

If a single person were to be credited with the creation of the Panama Canal, it would be Roosevelt. His ambition was to make the United States a global force, the "dominant power on the shores of the Pacific Ocean". As Secretary to the Navy, Governor of New York, and later President, Roosevelt campaigned vigorously for the canal, though for many years he believed that it should go through Nicaragua, not Panama. As President he connived in the creation of Panama, then overrode the wishes of Congress by giving supreme powers over the construction of the canal to one man, George Goethals.

Although three presidents were involved in the creation of the canal—Roosevelt, Taft and Wilson—it was Roosevelt who made the idea of it inspiring and inevitable. "The real builder of the Panama Canal was Theodore Roosevelt" according to Goethals. It could not have been more his own creation "if he had personally lifted every shovelful".

The gates at Gatún Locks during construction.

John Frank Stevens

Stevens was an engineer with an extraordinary record of success in constructing railways when Roosevelt appointed him to build the canal in 1905. In 1886 he had built a railway line 400 miles long through swamp and pine in upper Michigan, surviving disease, attacks by Indians and wolves, and the bitter cold of the North American winter.

When appointed to build the canal, Stevens inherited a mess; a year had passed, $128 million had been spent, but little achieved. There was still no plan, and no organization. Materials delivered to Panama were piled in heaps, and engineers were leaving as quickly as they could find passage on ships. Food was in short supply, disease rampant, and morale low.

Stevens stopped the work, and started to plan: he gave full support to proper sanitation, and reorganized the railways, essential to take away the spoil; he built a cold storage plant to provide decent food; he provided houses for his engineers and told them to send for their wives and families; and he built baseball fields and clubhouses, arranged concerts and created a healthy community.

Stevens lobbied hard for a canal with locks, and finally got his way. In 1906, he welcomed Roosevelt to the canal—a visit that swayed American opinion. Then, in February 1907, he wrote a long letter to Roosevelt complaining of exhaustion, of being constantly criticized, and describing the canal as "only a big ditch" whose utility had never been apparent to him. He asked Roosevelt for a rest, but the President took the letter as a resignation, and accepted it at once.

George Washington Goethals

Stevens' successor was George Goethals, a Lieutenant-Colonel in the Corps of Engineers. Roosevelt appointed him to chair the commission of seven which Congress had insisted be appointed, but made it clear that he was the boss.

When Goethals had examined the works in the company of Stevens, he paid tribute to Stevens's work: "there is nothing left for us to do but just . . . continue in the good work".

Goethals was a stiff, hard-working man who had few relaxations. He was tough, energetic, and not greatly loved, but he was a good picker of men and a good delegator. Every Sunday morning, between 7.30 and noon, any employee with a grievance or a complaint could come and see him. Goethals' Sunday morning sessions, in which he played a combination of father confessor and judge, were a totally new innovation in labour relations. They won him the support of the workforce, without

which the canal could not have been built.

Goethals was indomitable. When the walls of the Culebra Cut collapsed yet again, destroying months of work, Goethals hurried to the spot. "What do we do now?" he was asked. "Hell, dig it out again", replied Goethals. So they did, and went on doing until the canal was finally completed.

Dr William C. Gorgas

All the work of the engineers would have been to no avail without the work of Dr Gorgas, the man who brought the endemic diseases of Panama under control.

With the enthusiastic support of Stevens and Roosevelt, Gorgas eliminated mosquitoes, which he believed to be the carriers of the diseases. It might not have worked; knowledge was scanty, and Gorgas's opponents said that money was being wasted. But he was right. Within 18 months of his arrival, yellow fever was eradicated, and malaria was also coming under control.

Gorgas also had proper pavements and drains laid, hospitals built and sanitation provided. The country that had been the graveyard of De Lesseps' hopes was rendered tolerably healthy, in one of the most remarkable feats of public health ever accomplished.

Canals across the World

Credit for the first canal appears to rest with the Chinese, though there is evidence of a form of canal dating from about 4000 BC in Iraq; certainly the oldest recorded and still working waterway made by human hand is the Grand Canal linking Tianjin and Hangzhou, which was built between 485 BC and AD 283. It was also in China that the first pound lock was built, using two sets of gates to raise or lower boats between two levels.

The earliest canals in the West were short sections made to avoid obstacles impeding the navigation of rivers. Subsequently canals have been built to bypass whole rivers, to extend navigation from a river to a town, and to link rivers, lakes and seas. The longest canal joins the Volga River at Astrakhan with the Baltic Sea at Leningrad, a distance of 1,850 miles.

The Grand Canal
Like China's Great Wall, the Grand Canal was built in sections over many centuries. It uses parts of rivers, has been rebuilt, enlarged or rerouted, making it impossible to give a definitive length, but remodelling in the 13th century gave a route of 1,100 miles. Its main function was to help the collection of taxes paid in the form of rice grains. Today the canal carries 2,000-ton ships, but its usual traffic is smaller barges, like these in Suzhou.

The Suez Canal
A link between the Red and Mediterranean seas dates back to Herodotus (died 424 BC) who wrote of a canal from Suez to the Nile. Though Napoleon had a survey made for a canal, the idea lay dormant until 1833 when the eventual builder, Ferdinand de Lesseps, became involved. Doubts about the scheme delayed the start of work until 1860. Mechanical diggers removed about 100 million cubic feet of spoil, and the 100-mile-long canal was opened in 1869. This panorama shows Suez in the foreground and Port Said at the far end.

The Corinth Canal

A canal to link the Aegean and Ionian seas was begun under Emperor Nero in AD 67, but work ceased with his death. It was not until 1882 that it was resumed under a Hungarian engineer. The two ends, in the gulfs of Corinth and Aegina, were protected by breakwaters and the approaches dredged. The canal is a 4-mile cutting with an average depth of 190 feet.

The Ronquières Inclined Plane

Perhaps the greatest canal structure in the world is situated on the canal between Brussels and Charleroi. The mile-long incline was opened in 1968 and rises 223 feet to cut out 28 locks and a 1,149-yard tunnel. Barges are transported up and down in water-filled tanks that can accommodate 1,350-ton barges; each tank weighs 5,–5,700 tons.

Lines across a Nation

Fact file

One of the greatest railway construction feats

General Manager:
Cornelius Van Horne

Built: 1881–85

Length (Montreal–Vancouver): 2,920 miles

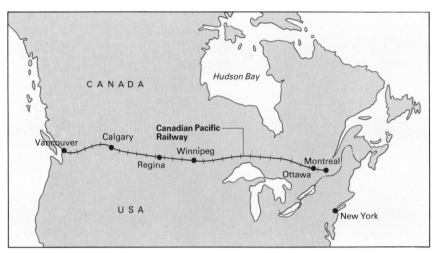

A tunnel near Rogers Pass which provided continual problems for the early operators of the CP. The death of 58 people due to a snowslide on the railway in 1910 prompted construction of the Connaught Tunnel, at 5 miles the longest double-track tunnel on the continent. From one vantage point in the Pass, Rogers could see 42 glaciers.

In 1871 the Conservative Prime Minister of Canada, John A. Macdonald, promised the colonists of British Columbia a railroad linking them to the east—within ten years. It was, said his Liberal opponent, Alexander Mackenzie, "an act of insane recklessness". Imprudent it may have been, but Macdonald had a dream: a British North America stretching from coast to coast, held together by a single line of steel. Without the railway, he feared, the nation could not be created, nor could British Columbia be persuaded to join the new confederation formed by Ontario, Quebec, New Brunswick, Nova Scotia and Manitoba.

The enterprise was epic in scale. Not only was the country enormous and thinly populated, but a transcontinental railway would have to run through some of the most inhospitable geography imaginable. At the time of Macdonald's pledge, large areas of the North West were hardly explored. Great mountain ranges and bottomless areas of muskeg swamp would somehow need to be surveyed and conquered if the promise were to come true. And how was a small nation of 3½ million people to find $100 million, the estimated cost of the railroad?

Things began badly, with a financial scandal and the fall of Macdonald's government. During the 1870s little was done except for the construction of branch lines and a tentative beginning on the main line at Fort William, Ontario. But by 1880 Macdonald was back in power and work began in earnest. By spring 1881 the financial problems had been ironed out and construction began at Portage la Prairie, west of Winnipeg. In November that year, a remarkable man, Cornelius Van Horne, was appointed General Manager of the Canadian Pacific Railway and charged with the task of building the railway. It was an inspired appointment, for Van Horne was a man of enormous energy, unquenchable optimism and considerable railway experience.

Van Horne swore that he would build 500 miles of track during 1882, and the whole railway in five years, half the time allowed by the government. He recruited 3,000 men and 4,000 horses and set to work across the prairie from Flat Creek to Fort Calgary. In April there were floods, in May snowstorms. By the end of June, hardly an inch of line had been laid.

People began to express doubts about the whole enterprise, but then began a construction blitz without parallel in railway history. From Winnipeg the line began to stretch across the country; every day, 65 flat trucks or box cars of supplies were dumped wherever the track ended, providing the raw materials for driving it onward. Ahead of them across the plains were the grading crews, flattening the ground with scrapers pulled by teams of horses. Accompanying them were bridging crews, throwing wooden bridges across rivers and streams, desperately trying to keep ahead of the track-layers.

By the arrival of winter, Van Horne had missed his target, but not by much. He had built 417 miles of track, 28 miles of sidings, and graded another 18 miles ready for the track-layers the following year. But where exactly was the track heading? That was an embarrassing question to ask in the winter of 1882. Van Horne and his men were moving as fast as they could across the prairie towards a double row of mountains—the Rocky Mountains and the Selkirks—without knowing a route over them.

The task of finding a route had been given to Major A.B. Rogers, a surveyor whose habit of

cursing at his workmen had given him the nickname of "Hell's Bells Rogers". He was honest, tough and ambitious. He was promised that if he could find a pass that would save the railway a 150-mile detour, he would be given a $5,000 bonus and have the pass named after him.

But first Rogers chose a route through the Rockies, selecting Kicking Horse Pass which was surveyed with extraordinary difficulty. Even the Indians avoided the Pass, regarding the treacherous gorge as too difficult for horses. An eastern pass was no good without a western exit from the mountains, but a route through the Selkirks was even harder to find. After several expeditions that courted death when supplies ran out, Rogers finally stumbled through spruce woods on to upland meadows from which a stream

flowed in opposite directions. A route that Rogers' detractors said was impossible had been found. The pass was given his name and the railway honoured their promise of a $5,000 cheque, though Rogers refused to cash it, preferring to hang it on the wall.

Kicking Horse was passable, but only just. As the railway inched across the map, money began to run low and economies were demanded. The contract with the government stipulated that the maximum gradient anywhere on the line was to be 2.2 per cent, or 116 feet in a mile. To achieve that through Kicking Horse would have involved building a tunnel 1,400 feet long, which would have delayed matters by a further year. Instead, a "temporary" line was built from the summit down into the valley at a gradient double

Lines across a Nation

A CP local train in 1900. Riding on the buffer beam at the front of a locomotive was considered one of the best ways to enjoy the spectacular scenery through which the CP passes in the west; in 1901 the party of the future King George V and Queen Mary rode in such style, well covered with travelling rugs, for a section near Glacier, British Columbia.

that permitted in the contract, and four times the maximum desirable. This was the notorious "Big Hill", 8 miles long, which for 25 years would terrify drivers and passengers alike.

The very first train that tried to descend the Big Hill ran away and fell into the river, killing three men. Safety diversions were installed and manned night and day. Every train had to stop at each one, and reset the points to the main line before going on. At the top of the hill, every passenger train stopped and had its brakes checked. A maximum speed of 6 mph was allowed, and guards jumped off from time to time to make sure the wheels were not locking. Going up was just as hard, needing four big engines for a 710-ton train of 11 passenger coaches. Not until 1909 was this notorious danger spot bypassed, by the expedient of driving two spiral tunnels into the mountain, by which the trains descended at gentler grades while curving round in a complete circle.

Construction of the line around the north edge of Lake Superior was another back-breaking assignment. The rock—granite striped with quartz—was hard, the winters tough, the summers made intolerable by flies. So much dynamite was needed to blast the line through that Van Horne established three factories, each

capable of turning out a ton a day. In the summer of 1884, 15,000 men were working on this stretch of line, costing $1.1 million a month in wages alone. In winter 300 dog teams, working non-stop, were needed to keep the men supplied.

By the beginning of 1885, the line was nearly finished, but so was the money to pay for it. Until the trains began to operate, there was no cash flow to offset the tremendous expenditure. The government refused to step in, and financial catastrophe threatened. At the end of March, just when all seemed lost, a rising in the North West by disaffected white settlers, backed by Indians, saved the situation. A force of 3,300 militiamen had to be sent west to quell the

The last spike is driven home at Eagle Pass by the eldest of the 4 CP directors present, Donald A. Smith, on 7 November 1885. The portly gentleman to his left is Van Horne, General Manager of the CPR, whose secret, he said, was "I eat all I can; I drink all I can; and I don't give a damn for anybody". The tall white-bearded man between them is Sir Sandford Fleming, who in 1862 first put before the government a considered plan to build a railway to the Pacific. Smith bent the first spike.

Stoney Creek Bridge on the long climb up to Rogers Pass being crossed by a transcontinental train with dome cars. Built in 1893, and reinforced in 1929, this steel arch of 336 feet replaced an earlier trestle viaduct dating from the opening of the line. Government cutbacks in 1990 abolished what has been described as the world's most spectacular rail journey.

The wheat plains of Manitoba proved easy country across which to build a railway. The gangs created an embankment 4 feet above the prairie, with ditches 20 yards wide on either side to prevent the line being blocked by snow.

Laying continuous welded rail near Lake Louise, Alberta, is a far cry from the basic and harsh conditions under which the line was built, though on good days 5 miles of track might be laid by the workmen. Their pay was between $1 and $1.50 for a 10-hour day. Their diet was salt pork, corned beef, molasses, beans, oatmeal, potatoes and tea. Lack of fruit gave them scurvy.

rebellion, and the unfinished railway was the only means of doing so swiftly enough to be effective. Van Horne promised to get them all to the North West within ten days, reasoning that no government could refuse aid to a railway that had helped it to crush a rebellion.

The journey was a nightmare, with the men travelling on flatcars or on horse-drawn sleighs along the unfinished sections, in bitter cold and snow. But they arrived and successfully quelled the revolt. Even so, the government would not help the railway until it came within a hair's breadth of collapse. On 10 July 1885, when Parliament was meeting to discuss aid, one of Canadian Pacific's creditors pressed for payment of a $400,000 debt. It was due for payment at 3 pm. At 2 pm, the House of Commons voted to provide more money. The railway was saved.

The final spike was driven at Eagle Pass on a dull November morning. All the leading figures in the railway, men who had come close to ruin to make it a reality, were on the spot to see it finished. There were cheers and the shrill whistle of a locomotive. Van Horne, called upon for a speech, made a short one: "All I can say is that the work has been done well in every way." Then the whistle sounded again, and a voice cried, "All aboard for the Pacific!"

Tracks through the Taiga

Few railways have been built under greater difficulties, or with more confusion, than the great line that runs 5,900 miles from Moscow to Vladivostock across the Siberian wastes. After constant false starts, and despite the urging of Tsar Alexander III, who declared in 1886, "It is time, it is high time!", actual building did not start until May 1891 at the eastern end, and a year later in the west. That it started at all was largely thanks to Sergius Witte, a railway enthusiast who was appointed Finance Minister and who, through brilliant financial stratagems, was able to repair Russia's crippled economy and provide the necessary money for the railway.

The building of the line was divided into sections, under the control of different engineers. The westernmost section, starting at Chelyabinsk, ran in a virtually straight line for 900 miles across level plains. But there were no trees to make sleepers, and all-out work was only possible for four months of the year.

Excavation was by pick and shovel, and to save money sleepers were more widely spaced than in Europe or North America, while the actual rails were made of much lighter steel. Ballast beneath the track was virtually non-existent; in many places the sleepers were laid on the earth. Progress, despite the problems, was rapid, with track being laid at the rate of $2\frac{1}{2}$ miles a day in summer conditions, and the first 500 miles of the western section was opened in September 1894. By August 1895 the Ob, one of Siberia's longest rivers, had been reached.

The crews built bridges as they went along, wooden structures over small rivers and streams, more substantial crossings of stone and steel for rivers like the Ob and the Yenisei. They built well, for many of these steel bridges survive today, despite the impact of thousands of tons of melting ice on the stone piers every spring. The cold claimed countless lives, for the gangs perched a hundred or so feet above the frozen rivers had little protection, and often allowed their bodies to become so chilled that they could no longer grasp the supports and fell on to the ice below. Most of the masons were Italians, who earned 100 roubles a month ($50).

Cast steel for the bridges came from the Urals, cement from St Petersburg, steel bearings from Warsaw, all brought along the newly built line at agonizingly slow speeds. Before bridges were built, the lines were sometimes laid bodily across the ice, frozen in place by constant douches of water. The few passengers alighted and walked across, while the driver gingerly tested the ice by driving on to it.

Fact file

The world's longest railway

Built: 1891–1904

Length: 5,900 miles

Duration of journey: 170 hours 5 minutes

Time zones traversed: 7

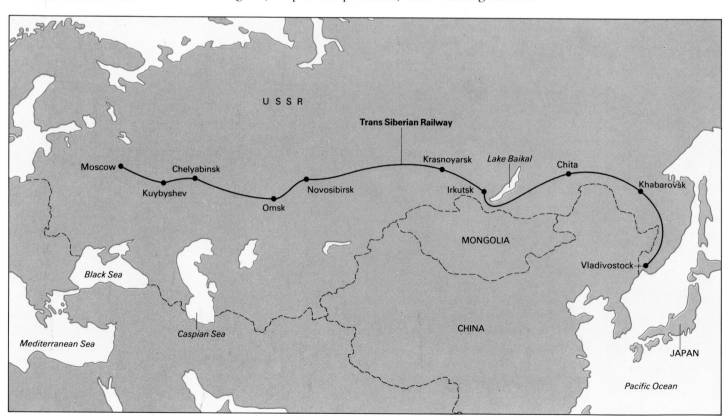

The seemingly endless forest through which much of the railway passes is seen here east of Krasnoyarsk, where it crosses the River Yenisei on a 1,010-yard viaduct of 6 spans. Beneath the emblem of Soviet railways (right) a plate bears the legend "Moscow to Vladivostock", which has only recently been opened to Western travellers.

Meanwhile another construction party had leap-frogged ahead to the mid-Siberian sector, where driving a railway through the virgin forest of the Siberian taiga was even more difficult. A passage 250 feet wide (to reduce the risk of fire from sparks) had to be cut through the forest, and rails laid on ground that was frozen until July, then turned into a boggy swamp. With 66,000 men at work by 1895 the mid-Siberian line was finished by mid-1898, that is in five years rather than the seven allowed.

The most difficult section of all still lay ahead, a 162-mile segment around the southern edge of Lake Baikal. Since construction would take some years, it was decided to take trains across the lake, the world's largest body of fresh water; ice-breaking train ferries were built on the River Tyne in England, taken apart, transported to Siberia and reassembled. Construction began in 1899, and was finished in feverish haste in 1904, after the war between Russia and Japan had broken out. So hasty was the work that the first train over the line derailed ten times. But on 25 September 1904 the line was opened, and it was possible for the first time to travel from the

Tracks through the Taiga

Atlantic shores of Europe to the eastern shores of Asia on uninterrupted rails. It had taken 13 years and 4 months, and it had cost $250 million.

Today the journey from one end of the railway to the other takes more than eight days, with 97 stops. Despite the achievement of building the line, and its economic and strategic importance, travel on the Trans-Siberian has never been as luxurious as on European trains such as the Orient Express, or especially exciting. The landscape unrolls day after day, unchanging and monotonous. The food is notorious; one writer remembers a week with nothing to eat but semolina and a kind of omelette. Nor are accidents unheard of. In June 1989 the world's worst-ever rail accident happened on the line 750 miles east of Moscow, when liquid gas from a leaking pipeline exploded as two packed trains were passing. More than 800 people died.

In the past 15 years the Soviet Union has built a second line across Siberia, the Baikal–Amur Mainline, or BAM. It runs for 2,000 miles across Siberia, passing north rather than south of Lake Baikal, and well to the north of the old line.

Construction workers *were recruited from Turkey, Persia and Italy, and even convicts from a prison near Irkutsk were requisitioned. They received a promise that 8 months' labouring was equal to a year off their sentence.*

This cutting near Lake Baikal *indicates the reason for the section of line around the lake being the last to be completed. It passes through 40 tunnels and numerous rock-hewn cuttings.*

The Russian Imperial arms of Nicholas II emblazoned over the entrance to the viaduct across the Volga between Syzran and Kuybyshev.

Ob' station, to the west of Novosibirsk, a small, ornate station with decorative bargeboards. The poor standard of engineering work on the railway meant that early speeds were slow: the Siberian Express averaged only 20 mph.

Passenger accommodation in the early years of the railway ranged from a sub-European standard train de luxe to converted goods vans (left), although use of the latter was usually for convicts, immigrant trains of compulsory exiles to Siberia, and even troops.

A Concrete Triumph

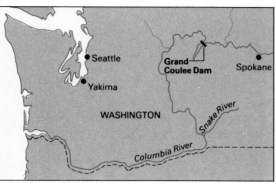

USA

Seattle
Yakima
WASHINGTON
Grand Coulee Dam
Spokane
Snake River
Columbia River

Fact file

When planned, the largest hydro-electric scheme in the world

Constructors: Mason-Walsh-Atkinson-Kier Co

Built: 1933–

Material: Concrete

Height: 550 feet

Length: 4,173 feet

The world's largest concrete dam—and the largest concrete structure in the world—lies on the Columbia River in the State of Washington, in the north-west United States. It is also one of the largest hydro-electric power plants in the world, and its huge irrigation pumps are big enough to pump dry most of the rivers in the USA. Its construction in an isolated and thinly populated area of the US during the Depression years was one of the great achievements of the Work Projects Administration, President Franklin Roosevelt's bid to restore prosperity and jobs to a nation in trouble.

The dam had two purposes, three if the provision of jobs is included. The principal one was to supply irrigation water to more than a million acres of desert land—known as the coulee country—in central Washington, good soil that needed only water to make it fertile. It used to be said that a jackrabbit had to carry his lunch and a canteen of water just to cross the coulee country, and the area was littered with empty farmsteads, broken windmills, and abandoned equipment. The second purpose was to generate electricity, some of which would be used to drive the irrigation pumps.

Geology had created the setting for the dam. Millions of years ago a glacier advancing southwards from Canada had cut off the flow of the Columbia River and forced it into a new channel. This channel was eventually eroded by the river to a depth of 900 feet, 5 miles wide, and 50 miles long, before the glacier retreated and allowed the river to return to its original course. The channel that was left was the Grand Coulee, dry along its entire length.

The Grand Coulee project envisaged a huge concrete dam across the new course of the river, creating a lake 151 miles long stretching back towards Canada, and two smaller earth dams across the Grand Coulee itself, to turn it into a storage reservoir for irrigation. Because the glacier had raised the level of the river, the storage reservoir in the Grand Coulee lies about 300 feet above the high-water mark in the lake below, so pumps are needed to carry the water up to it. From here, the water is distributed through canals to the high and arid lands on the plateau of the coulee country.

The main dam is of prodigious size, 4,173 feet long and at 550 feet as tall as a 46-storey building. It contains 10,979,641 cubic yards of concrete, and raised the level of the river by 350 feet. It depends entirely on its mass to resist the pressure of water, the river being too wide for an arched dam. Preliminary engineering work began in 1933, and the first contract was let at the end of that year.

In order to build the dam on sound foundations, temporary cofferdams made of steel piling and timber were constructed to narrow the width of the river and expose the bedrock below. Two U-shaped cofferdams were built, one on either side of the river, leaving a gap just 500 feet wide for the river to flow through. Water was pumped from the areas inside the cofferdams, and the rock exposed. Once the areas were dry, the concrete dam was constructed outward from each bank toward the middle, leaving some low areas as spillways. Then two more cofferdams were built above and below the dam, diverting the river over the spillways so that the central part of the river could be pumped dry and the last 500 feet of the concrete laid.

The concrete was poured as a series of columns 50 feet by 50 feet which extended from the bedrock to the full height of the dam. The columns grew 5 feet at a time, with 72 hours between each successive pour to give the concrete time to begin curing.

When curing, concrete generates heat in a chemical reaction. If this heat is not removed, a huge concrete structure will increase in temperature over a period of months, and expand. When

curing is complete, and the temperature falls, it will shrink, creating cracks. To prevent this happening, special cooling pipes made of thin wall steel tubing an inch thick were set into the concrete as it was poured, and cooling water pumped through.

When the columns of concrete were finally cool and set, the small gaps between them caused by shrinkage were filled by pumping in a cement and water grout through a network of pipes embedded into the concrete as it was poured. The shrinkage between each block was tiny—only $\frac{3}{32}$ of an inch—but over the length of the dam that added up to about 8 inches. The grout sealed the blocks, making a watertight dam.

A novel problem emerged during the building of the dam, and was solved in an unusual way. At the east end of the dam, the bedrock was no sooner exposed than it was covered by a huge volume of plastic clay, which kept creeping forward and defeated all attempts to stop it. Timber and concrete obstacles were set up, but to no avail. The volume of clay involved was 200,000 cubic yards, so removing it would have been time-consuming and expensive. Finally the engineers had the idea of freezing the leading

*The **Grand Coulee Dam** is the main dam; the more recent Forebay Dam, which is still to be fully completed, is the shorter, angled dam on the left. Behind the dam the lake extends for 151 miles toward British Columbia, with an average width of 4,000 feet and a depth of 375 feet. The lake is a new haven for wildlife.*

A Concrete Triumph

Excavated spoil from the east side was carried 4,000 feet across the river by conveyor belt (top) to Rattlesnake Canyon where 13 million cubic yards of spoil were deposited. Cement was brought from 5 plants in Washington state and stored in steel silos before being combined with sand and gravel in mixers 100 feet high (above).

edge of the clay to form a dam that would hold the rest back. A 3-mile length of pipe was placed in an arc in the toe of the sloping mass of clay, and brine at a temperature of zero degrees Fahrenheit circulated through it. This froze the front edge of the clay into the shape of an arch 20 feet thick, 45 feet deep and 110 feet long.

Between August 1936 and April 1937 an ice plant kept the clay frozen while the bedrock was prepared and the dam constructed until it was above the level of the clay. After that, the ice plant was switched off, and the clay once more allowed to move. It had cost $35,000, but saved many times that much.

The lake created by the dam holds water amounting to 20,000 gallons for every citizen of the US, but such is the flow of the Columbia River that it could fill the lake in two months, or in one month in the flood season. At each side of the river, hydro-electric plants were built, initially with a total output of 1,920 megawatts. Power from these is used to drive 12 pumps located on the west side of the river behind the dam. The capacity of each of these pumps, 1,600 cubic feet per second, is enough to irrigate 120,000 acres of land.

The pumps send water through conduits 13 feet in diameter to the higher reservoir, built in the upper Grand Coulee. This was created by building two earth dams about 100 feet high, one

The first phase of the Forebay Dam under construction in 1971 at the west end of the Grand Coulee Dam (right). A close-up shows the construction of the 6 penstocks (below)—tubes that feed water into the turbines. Their 40-foot diameter is over twice the size of those in the original power plant. Each penstock is made up of cylindrical sections, or "cans", which are lowered down rails and welded in place.

Two trestles, each 3,000 feet long (above), were built to enable concrete to be placed in the columns. Cranes with a reach of over 115 feet could travel along the trestle, picking up bottom-emptying buckets that were brought on to the trestle by railway from the concrete mixers. Concrete could be placed at 1 cubic yard every 5½ seconds.

about 2 miles from the Grand Coulee Dam and another near Coulee City. Between these two dams a reservoir 27 miles long was created, and filled with water pumped 300 feet up from the big reservoir below. From here the water flows 10 miles to the heads of two canals, the 150-mile east-side canal and the 100-mile west-side canal, from which it is distributed to farms through shorter lateral channels.

Since 1970, a new scheme to increase the output of the hydro-electric plants has been under way. This has involved removing some 250 feet of the existing dam at the east end, and building a new dam, joined to it and angled away downstream. The new turbines are built into this

dam, a simpler task than taking apart the old dam to replace the turbines installed in the 1930s. To remove the end of the original dam, a cofferdam was first built to isolate it, then it was carefully demolished block by block by toppling the blocks over downstream using dynamite.

Simultaneously, penstocks—huge tubes to carry the water to the turbines—were being constructed downstream of where the new section of dam was to go. The ultimate output of the plant, when all phases of the contract are complete, will be 10,080 megawatts. Only one other plant in the world, the Itaipu power station on the Parana River on the borders between Brazil and Paraguay, exceeds this.

Harnessing the Power of Water

The earliest known dam was a series of stone-faced earth dams at Jawa in Jordan, dated to *c*3200 BC. Earthen irrigation dams in the Tigris and Euphrates valleys were followed by the first known rock-filled dam, near Homs in Syria, built *c*1300 BC. The skills of dam building spread to India, Sri Lanka and Japan. The first known arch dam, achieving strength through the same principle as an arch bridge, was built on the Turkish/Syrian border during the reign of Justinian I (AD 527–65).

In the twentieth century, the generation of hydroelectric power has become the principal justification for many dams, a development pioneered by Sir William Armstrong whose Northumberland home was the first in the world to be lit by this method.

Kielder Dam, England
The largest dam in Britain, Kielder's embankment measures 3,740 feet in length and contains almost 7 million cubic yards of earth. Loss of countryside through flooding is often the cause of opposition to dams, but Kielder has won praise for the way it has been landscaped. The reservoir of 2,684 acres supplies water to a large part of north-east England.

Tucurui Dam, Brazil
The capacity for dams to alter beyond recognition substantial areas of land is illustrated by this $4 billion dam on the Tocantins River. The dam, of 84 million cubic yards, turned the river into a chain of lakes 1,180 miles long. Hydroelectricity has been so widely adopted that about 20 percent of the world's electricity is derived from water-driven turbines. Much of the work on turbines was done by the British engineer Charles Parsons, who invented the steam turbine in 1884.

Itaipu Dam, Brazil
Built at a cost of $11 billion on the Parana River on the Paraguay/ Brazil border, Itaipu power station began generating power in October 1984 and will produce 13,320 megawatts from the 18 turbines in the dam. It is the world's most powerful dam, although a much more powerful project is planned for the Tunguska River in the USSR. It is hydroelectric schemes in Brazil that have aroused the greatest opposition to the international bank loans that finance such projects, seen by many as ecological disasters. One of the biggest schemes in Brazil, Plan 2010, envisaged constructing 136 dams to supply Brazil's energy needs for the next 2 decades. It would have entailed the flooding of an area the size of the United Kingdom and displaced 250,000 Indians from their traditional rainforest.

Taming the Seas

Fact file

The world's largest sea barrier

Built: 1958–86

Materials: Prestressed concrete and steel

Length: Eastern Scheldt, 2,743 yards

The Netherlands are not known as the Low Countries for nothing. The Dutch people have struggled against flooding for nine centuries, pioneering the building of dykes, dams and canals to tame the sea and to recover new land for agriculture. At the same time, they have used their easy access to the sea to become a major shipping and trading nation, with the port of Rotterdam holding its position as Europe's busiest. In October 1986 Queen Beatrix inaugurated the biggest and most advanced sea barrier anywhere in the world, the culmination of a plan that had been in progress for almost 30 years.

The Dutch Delta Plan, unlike the damming of the Zuider Zee, was not designed to create new land from the sea. Its object has always been to reduce the danger of catastrophic flooding, which in the past has periodically swept aside the dykes and engulfed large areas of the country. The last great disaster was on the night of 31 January–1 February 1953, when huge areas were inundated by the combination of a spring tide and a severe north-westerly gale. Hundreds of dykes were swept away, 395,000 acres of land flooded, and 1,835 people lost their lives in the worst flood ever to hit the Netherlands.

Plans had already been started for a system of dams, but the disaster created much more urgency. In 1958 Parliament passed the Delta Act, a bold attempt to remodel the whole coast of south-west Netherlands, preventing any future danger of floods, while keeping open the waterways giving access to the ports of Rotterdam and Antwerp.

The plan envisaged a series of dams and surge barriers, some with locks and discharge sluices, to force the saltwater back toward the sea, prevent flooding, and improve fresh-water management in the country. The plan was put into action step by step, starting with the smaller projects and gradually moving to more difficult ones as experience was gained. It involved five primary dams, five secondary dams, the strengthening of dykes along the New Waterway which leads to Rotterdam and the Western Scheldt which leads to Antwerp, and the building of two major bridges. The first primary dam to be built, the Veerse Gat, sealed off an estuary with a tidal volume of 6,200 million cubic feet; the last, the Eastern Scheldt, had a tidal volume of 77,500 million cubic feet.

The height of the dams was set at a level approximately 3 feet higher than the level reached in the 1953 flood. The chances that such a level will be exceeded were calculated at less than one in 10,000 for the most important economic areas—giving a 1 percent chance of a flood topping the barriers in any 100 years— and one in 4,000 for the rest. Among the greatest problems faced in building the dams were scouring by the sea, which tends to wash away the foundations, and devising ways for closing the final gap in the dam when it was virtually complete. At the Haringvliet Barrage, for example, huge underwater "aprons" had to be built on either side of the dam, consisting of a nylon mattress with layers of graded gravel and rocks on top of it. The barrage had to have gates strong enough to resist the force of the sea, spanning sluiceways big enough to pass ice in winter.

Two methods were used for closing the final gaps in the dams. One was to build a series of prefabricated concrete caissons, which could be placed in position at slack water, gradually narrowing the gap until the final caisson was placed. Once the water had been dammed in this way, the rest of the dam could be built around the caissons. This was the method used at the Veerse Gat Dam. The alternative was to build the dam outward from either shore, and then run a cableway across between the two ends. The cableway carried cars loaded with stone which

Schaar construction dock (above), in which the prefabricated piers of the Eastern Scheldt barrier were built. The dock is divided into 4 by dykes; as soon as all the piers in a compartment were complete, it was flooded to enable a vessel to lift them, one by one, and drop them in position. The Delta Plan was devised to prevent inundation of low-lying land (right).

were hauled out to the middle and their load dumped into the gap, gradually filling the breach until the stone emerged from the water. This method was used at Grevelingen Dam.

The last and biggest task in the Delta Plan was the damming of the Eastern Scheldt, a huge body of tidal water, with a barrier more than 5 miles long. Final closure was planned for 1978. But a strong campaign was launched in favour of keeping the Eastern Scheldt open, in order to preserve the natural environment. Sealed off, it would have lost its role as a nursery for North Sea fish, and its attraction to the seabirds that flock to the sandbanks at low tide. The growing of shellfish would also have been eliminated if the Eastern Scheldt became an inland lake cut off

Taming the Seas

from the sea. Against that, those anxious for the greatest possible safety from flooding, and the improvement of agriculture, wanted the dam.

After years of argument the Dutch government ordered a study to see if it were possible to modify the dam, providing instead a barrage that would stay open all the time, except when a storm surge was expected, when gates would be closed to prevent flooding. The danger was that such a plan would retain as a permanent feature of the barrage the most dangerous stage in the closure operation, which many engineers regarded as foolhardy. Despite that, the Dutch government decided to bow to the pressure from the environmentalists, and turn the dam into a barrage, despite the engineering difficulties.

The design of the barrage consists of 65 prefabricated concrete piers between which 62 sliding steel gates have been installed. The gates are more than 17 feet thick and 130 feet wide. With the gates in the raised position, the tidal range behind the barrier would remain three-quarters of what it was originally, thus preserving the natural environment.

The barrage is built across the three main tidal channels of the Eastern Scheldt, with the rest of the crossing consisting of a dam. When the gates are closed, the forces exerted on the gates and the piers are enormous. The foundations must be designed so that these forces do not shift the piers, which would cause jamming of the gates.

The piers are placed on the sea bed on top of foundation mattresses, the purpose of which is to absorb changing water pressure in the subsoil so that the fine sand under the mattress is not washed away, thus weakening the foundations. Slight variations in the level of the mattresses were compensated for by additional concrete mattresses of varying thickness, before the piers were placed on top and sunk into place. They are not attached to the sea bed by any piling, but remain in place simply by virtue of their weight.

Around the base of each pier a sill was constructed of graded layers of stone, each layer becoming larger the closer it is to the surface. The bigger stones in the upper layers prevent the smaller ones below being swept away. The top layer consists of basalt rocks weighing between 6 and 10 tons, designed to ensure that if a gate should fail to close the current rushing through it will not carry away the stone and endanger the barrier.

The piers are linked by two sets of sills, made

Foundation mattresses (above) for the piers were made in a special plant and consisted of successive layers of sand, fine gravel and coarse gravel on a plastic base. Each more than 650 feet long, 140 feet broad and 1 foot thick, they were laid by a special rig aboard 2 ships, followed by a second smaller mattress laid on top.

of concrete. The lower sills each weigh 2,500 tons and link the piers underwater. The smaller upper sills weigh 1,100 tons and form the upper edge of the opening through which the tide passes when the gates are open.

The operation of the gates is tested at least once a month, and it takes about an hour to close or open them all. If they are closed too quickly, it can affect the wave movements inside the barrier. While emergency closure to prevent flooding is a rare occurrence, the gates are used to fine tune the amount of water flowing in and out of the Eastern Scheldt.

The barrier was finally completed in 1986, the largest Dutch public works project since the Second World War, costing $2.4 billion. In the spring of 1990 it successfully staved off a flood of potentially disastrous proportions.

The 65 prefabricated piers (left) *were built in the Schaar construction dock. Each pier is the height of a 12-storey building, weighs 18,000 tons and took 1½ years to complete. Work began on a new pier every 2 weeks so that their installation would be an uninterrupted process. In 4 years, 589,000 cubic yards of concrete were used to make the piers.*

Placing the 18,000-ton piers in water up to 100 feet in depth demanded accuracy to a few inches. A U-shaped vessel, the Ostrea (right), lifted a pier, transported it to the site and manoeuvred it into position with the help of its 4 rudder propellers, 2 in the bow, 2 in the stern. Another vessel held the Ostrea steady while the pier was lowered into place.

The steel gates (right) *are more than 17 feet thick and 130 feet wide, the precise dimensions being determined only after the piers were in place. The height of the gates varies from 77 feet to 155 feet, depending on their position in the channel. Their weight ranges from 300 to 500 tons.*

The completed storm surge barrier showing the vertical hydraulic cylinders protruding above the capping units. Each gate is opened and closed by 2 cylinders operated from the central control building.

Conquest of the Alps

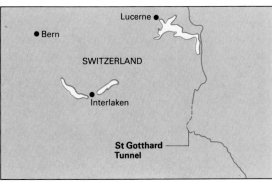

Fact file

The world's longest road tunnel

Engineer, rail: Louis Favre

Built: rail 1872–82 road 1970–80

Length: rail $9\frac{1}{4}$ miles road $10\frac{1}{2}$ miles

From Göschenen in the Swiss canton of Uri to Airolo, in Italy, is just 10 miles through the road tunnel under the Alps which was opened in 1980. But providing a route between the two places by road or rail had taxed the ingenuity of engineers for hundreds of years. Today it is possible to drive all the way from Hamburg in West Germany to Reggio di Calabria in the toe of Italy without ever leaving a motorway, thanks to the tunnel—the longest road tunnel in the world.

The St Gotthard Pass has always been important because of its position on a direct line joining Milan with the Rhine valley. By Alpine standards it is not particularly high—6,936 feet above sea level, but it was never easy going, due to an awkward obstacle on the Swiss side, the narrow and precipitous Schöllenen gorge above Göschenen. At the beginning of the thirteenth century, unknown engineers managed to span this gorge with a narrow wooden bridge some 100 feet above the Reuss River. The approach to the bridge, along vertical cliffs of stone, was formed by a wooden walkway 80 yards long, fastened to the rock with chains. It was a major achievement of medieval technology, known locally as the "Devil's Bridge" because only the Devil would have had the ingenuity to create it.

In 1595 the wooden bridge was replaced with a stone arch, and in 1707 the first Alpine tunnel, the Urnerloch, was driven through the mountainside to replace the walkway. It was 80 yards long, and excavated by Petro Morettini of Ticino. It was just 12 feet by 10 feet, too narrow for carriages, although in 1775 an English mineralogist named Greville drove a light chaise through it, becoming the first person to cross the St Gotthard by vehicle. In 1830 the gallery was widened to take full-size carriages.

By then it had been the setting for an extraordinary clash in 1799 between a force of 21,000 Russian troops led by General Suvarov and the armies of Revolutionary France. In a 12-hour battle fought on the Italian side of the St Gotthard Pass, Suvarov defeated the French, but in retreating they destroyed the Devil's Bridge and left the Urnerloch defended by a rearguard. In a bitter battle, many Russians died as they tried to dislodge the French, succeeding finally by finding a ford across the river. The bridge was repaired and Suvarov led his men across.

Between 1818 and 1830, the road across St Gotthard was greatly improved by the Uri canton, which was almost ruined by the expense of the operation. In those days the passes did not close in winter, but were kept open by hardy *cantonniers* who went out after snowfalls with ploughs, pulled by oxen, to cut a narrow passage through the drifts. Passengers were transferred from carriages to horse-drawn sleighs, enveloped in fur coats and blankets, and taken over the top through the paths the oxen had carved out. Special passing places were provided, in which the horse going up had the right of way. At the summit of the pass was a hospice, where travellers could be thawed out and fed.

By 1850, the railways could bring travellers to either side of the great Alpine passes, but it still required carriages, horses, and sleighs to get them over. During the 1860s the first Alpine tunnel, under Mont Cenis, was successfully dug, and in 1871 Germany, Italy and Switzerland signed an agreement to subsidize the digging of a tunnel through the St Gotthard massif. The man who won the contract was Louis Favre of Genoa, and it destroyed him. Tough penalty clauses for late completion of the tunnel threw his company into bankruptcy, and Favre himself died a broken man before the tunnel was finally opened in May 1882 after ten years of work.

A principal trouble was water, which poured into the workings with the force of a fire hose, and forced workers to excavate the tunnel up to their knees in water. Dynamite was no sooner placed in holes to blast the rock than it was

The approaches to the Gotthard railway tunnel from both directions required engineering as impressive as the tunnel itself. From the north, on grades as steep as 2.6% or 1 in $37\frac{1}{2}$, the line describes a complete circle within the Pfaffensprung Tunnel, followed by 2 other tunnels, besides crossing numerous viaducts. The southern approach (above) was equally dramatic, with 2 spiral tunnels above this crossing of the Ticino River in the Piottino Ravine.

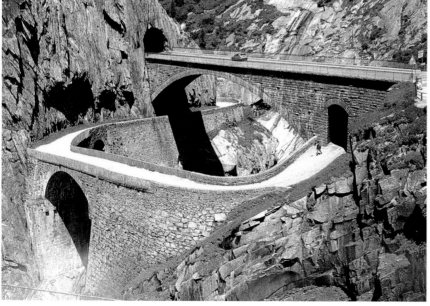

washed away as a yellow sludge. The temperature in the tunnel was tropical and disease flourished, with men and horses dying or being forced to give up work. Air was pumped to the working face by compressors, and used to drive the drills and provide air to breath; but the compressors were not up to the job, and miners found themselves gasping for air.

The greatest problem of all was the collapse of the roof in a section of tunnel $1\frac{1}{4}$ miles from the Swiss end. The tunnel had been driven at this point through unstable gypsum and feldspar, which in contact with the moist air began to liquefy, exerting such huge force on the tunnel linings that it crushed them. It took two years to solve the problem, which was achieved by building a massive granite wall more than 8 feet thick, carrying an arch 4 feet 8 inches deep, which proved strong enough finally to hold back

Conquest of the Alps

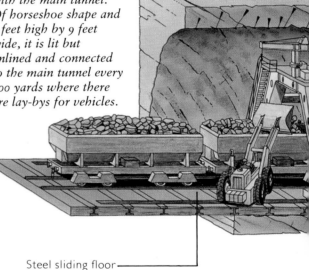

The safety tunnel (above) runs parallel with the main tunnel. Of horseshoe shape and 8 feet high by 9 feet wide, it is lit but unlined and connected to the main tunnel every 300 yards where there are lay-bys for vehicles.

Steel sliding floor

Full-face excavation was carried out by drilling jumbos which each weighed 36½ tons and had 5 hammer drills, powered by compressed air. Smaller drills (above) were used to make holes for bolts to secure the protective wire netting above the working area until steel supports could be installed.

the sticky mass. It was an expensive delay.

The final cost of the tunnel was 57.6 million francs (£2.3 million) which was 14.7 million francs (£590,000) more than Favre had bid. In today's circumstances, and given the huge difficulties overcome, that may seem a modest overspend but the railway company insisted his firm bear the loss—a judgement upheld by the courts—and that he also pay forfeit money of £230,000 for late completion. Favre was already dead, but these harsh conditions ensured that his company disappeared also. He was not the only victim: the St Gotthard Tunnel cost the lives of 310 workmen, and incapacitated another 877.

Today the same mountain has been pierced by a second tunnel, which carries the road. That, too, proved a tough undertaking. Work began in 1969, and the tunnel was finally opened 11 years later, in September 1980. Under pressure from Swiss motoring organizations, a safety tunnel was built parallel to the main tunnel and 100 feet away to provide an escape route in the event of fire. The two tunnels follow a curving path through the mountain, partly to provide shorter ventilation tunnels accessible from the existing pass, and partly to avoid tricky rock. In addition, it was felt that driving around a gentle curve for 10 miles, together with some changes of gradient, would be less tiring and potentially hazar-

dous than holding a perfectly straight course.

Drilling of the tunnel began from both ends on 5 May 1970, using a conventional tunnelling method in which rock was broken up by explosive charges and then removed by road or rail vehicles from the working face. The safety tunnel and the four ventilation tunnels were excavated at the same time, the safety tunnel running slightly ahead to provide warning of poor ground conditions. The main tunnel is 25 feet wide, allowing for a single carriageway in each direction, and 14 feet 6 inches high.

One innovation in the drilling of the tunnel was the use of a "sliding floor", made of steel, and more than 250 yards long. As excavation proceeded, the floor was moved forward, providing a sound surface for trucks coming to take away the rock, and a platform from which to build the lining to support the tunnel roof. To ensure that the tunnel ran in the right direction, a system of nine laser beams was used, mounted at

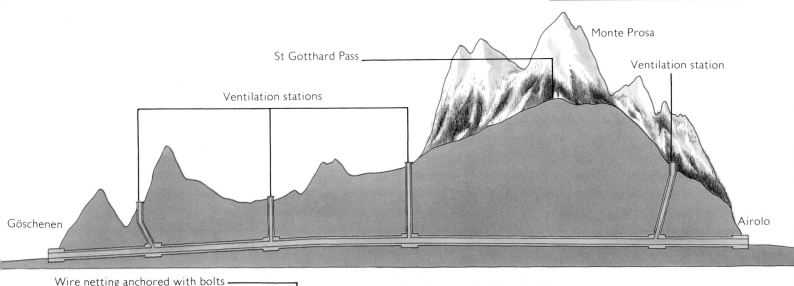

St Gotthard Pass

Monte Prosa

Ventilation station

Ventilation stations

Göschenen

Airolo

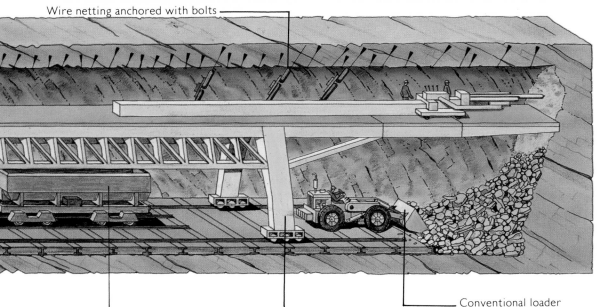

Wire netting anchored with bolts

Six ventilation plants
(above), *together
consuming up to 24
megawatts, work to
remove potentially
lethal exhaust fumes. At
full power they are
capable of changing the
air in the tunnel within
6 minutes, but when
traffic is light, a
computer controls fans
to avoid wasting power.*

Railway truck for spoil removal

Conventional loader

Supporting platform and cantilever arm

the opening of the tunnel and pointing forward to mark out the profile for the excavation. After every advance of 350 yards, they were moved nearer the face and aimed in the right direction for tunnelling to proceed. At one point, about ½ mile from the northern portal, the tunnel passes directly underneath the rail tunnel, with a gap of only 16 feet of rock, so care was needed with blasting to avoid damage to the older tunnel.

About 730 workers were employed, half at each end, and they joined up in March 1976. In spite of the safety precautions taken, 19 died in accidents during the excavation. The tunnel they built is filled with special systems to ensure the safety of its users. Lighting is continuous, with every tenth light connected to a separate, independent power supply.

Within the tunnel, and on the approaches to it, there are signalling systems to control the flow of traffic, which has to filter down from the two-lane carriageways on the approach roads to the

A protective platform
(above), *32 yards long
and 5½ yards high with a
cantilever arm of 15
yards, enabled
simultaneous work on
the excavated arch and
removal of blasted rock.*

single carriageways in the tunnel. Along the full length of the tunnel there are shelters on the east side which lead directly to the safety tunnel, and on the west side are niches, with fire extinguishers and emergency telephones, within which people can take cover in the event of fire or accident. The tunnel is also fitted with fire alarm systems and TV monitors, while radio systems enable drivers to listen to their radios as they go through, and hear any emergency warnings.

Like Favre, the tunnellers met difficult rock and an unexpected amount of water under the St Gotthard massif, delaying the opening of the tunnel for three years. Its final cost was 690 million Swiss francs (£175 million) which was more than twice the original budget. Inflation accounted for half the increase, and the difficult rock for a quarter. Some extra money was also spent on preparing the way for a future second tunnel, when the capacity of the present one (1,800 vehicles per hour) is reached.

The World's Greatest Highways

The first roads were little more than tracks that developed into trade routes, of which the most famous were the silk routes between Persia and China. The first surfaced roads were made by the Egyptians, who built polished stone causeways to facilitate the carriage of stone blocks for the pyramids. However, it is the Romans who are renowned for their skills in creating straight roads across their empire.

The development of wheeled vehicles called for better roads, leading from the work of British engineers on turnpike roads, through the auto-bahns and autostradas of the 1930s, to today's multi-lane highways. Concern over their environmental impact is now forcing a reappraisal of road schemes and traffic growth.

The Appian Way
Completed between the Porta San Sebastiano in Rome and Capua in 312 BC by the censor Appius Claudius, the Via Appia (below) was the most important of the consular roads. Later extended to Benevento and Brindisi, it was surfaced with huge polygonal blocks of basaltic lava, and lined for the first part with family tombs and temples.

Los Angeles Interchange
Of the world's cities, Los Angeles is probably the most wedded to the automobile: a million cars a day carry 3.3 million commuters along 725 miles of generally clogged freeway, producing so much poisonous smog that radical solutions are being considered to try to stop the city choking to death.

The Pan-American Highway

The world's longest road begins in Texas (several places contend for the honour) and runs for over 15,500 miles through Panama City to Valparaiso in Chile; from there it turns east to cross the Andes, seen here near Arequipa in Peru, to Buenos Aires. Short gaps exist in Central America, and there is an extension to Brasilia.

Darby's Metal Masterpiece

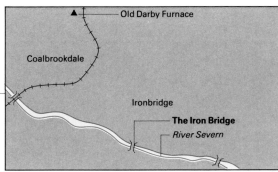

Old Darby Furnace

Coalbrookdale

Ironbridge

The Iron Bridge
River Severn

ENGLAND

Fact file

The world's first iron bridge, built over the River Severn

Designer: Thomas Farnolls Pritchard

Built: 1777–79

Materials: Cast iron and stone

Length: Span of 100 feet 6 inches

Weight of iron: 378 tons 15 cwt

"From Coalport to the Ironbridge, two miles, the river passes through the most extraordinary district in the world", wrote a Shrewsbury man, Charles Hulbert, at the end of the eighteenth century. Everywhere, he said, were iron works, brick works, boat-building establishments, retail stores, inns and houses.

Hulbert was describing the first district in England, and the world, to feel the impact of the Industrial Revolution. It was a change so radical and complete that it has dominated human lives ever since. And the most eloquent symbol of that revolution is the bridge Hulbert mentions, a bridge made of iron across the River Severn. From the beginning the bridge was a source of wonder: the dramatist and song-writer Charles Dibdin wrote that "though it seems like network wrought in iron, it will apparently be uninjured for ages". Uninjured, it stands today.

The idea for the bridge came from a Shrewsbury architect, Thomas Farnolls Pritchard. The bridge would replace a ferry across the gorge of the Severn between Madeley and Broseley, reducing the delays and inconveniences, particularly in the winter, caused by the poor boat services. Why Pritchard opted for cast iron is less clear, although in the parliamentary bill giving assent to the bridge, passed in the spring of 1776, the options were widened to include structures of "cast iron, stone, brick or timber". The petition to Parliament merely claimed that it would be "of public utility" if the bridge were to be constructed of cast iron—presumably on grounds of longevity and strength, and in order to demonstrate the possibilities of the material.

By the summer of 1776, indeed, the trustees were split between the radicals, led by the ironmaster Abraham Darby III, who favoured the cast-iron structure, and the conservatives who preferred a more conventional solution. Fortunately Darby, though outnumbered, held the majority of the shares and was able to get his way. Pritchard and Darby estimated, between

them, that the bridge would be built for £3,200, of which £2,100 was to go on more than 300 tons of cast iron and £500 on dressed stone. These figures turned out to be a considerable underestimate, and during the building of the bridge financial catastrophe was never far away.

Darby, chosen by Pritchard to be the builder of the bridge, was the third member of his family to carry the same name. His grandfather, Abraham Darby I, devised in 1709 a method of making iron in a blast furnace using coke rather than charcoal as the source of carbon. The discovery, although of enormous long-term importance, was taken up only slowly by other iron makers. The reason was that for the time being there was no great shortage of charcoal, and coke worked well only if the ore and the coal were carefully chosen. But by 1755 his son Abraham Darby II, also working in Coalbrookdale, had built a blast furnace using coke that was fully competitive with charcoal and produced cast iron of high quality. It was his son, the third in the Darby dynasty, who was to apply this material to the building of the bridge.

The first designs by Pritchard were for a single-span bridge of 120 feet, with four sets of curved iron ribs, each 9 inches by 6 inches in section. But by July 1777 the span had been reduced to 90 feet, and the design changed slightly. It was subsequently increased once more, to 100 feet 6 inches, to accommodate a towpath along the banks of the Severn, and this is the bridge that was eventually built. On 21 December 1777, when work had hardly begun, Pritchard died, leaving the task to Darby.

The biggest castings in the bridge are the main ribs, each of which weighs $5\frac{3}{4}$ tons. At the time, the blast furnaces of Coalbrookdale produced little more than 2 tons of iron at a time, so the ribs could not have been poured into moulds directly from the blast furnace. The chances are that a special remelting furnace was set up on the banks of the river, and used to melt iron that had been made earlier in the blast furnace, before it was poured into sand moulds.

The advantage would have been that the heavy and fragile castings would not then have had to be manhandled for just over a mile from the Coalbrookdale foundry to the river bank. Though strong in compression, cast iron is a brittle material that required careful handling until it was safely in place. Once erected, the design was such that the forces exerted on the ribs were predominantly compressive.

Considering the contemporary interest in the structure, accounts of its construction are scanty. It appears that most of the work of raising the ribs was carried out in about six weeks in the summer of 1779, beginning with the raising and securing of the first pair of ribs on 1 and 2 July. Some 25 to 30 men were employed on the job. Darby's accounts record the purchase of large amounts of timber, which were used to build a scaffold from which the ribs were suspended on ropes and lowered until they joined in the middle.

In mid-August, the account books record the spending of £6 on ale, which we may guess was to celebrate the completion of the fitting of the ribs. By late November, the same books show the removal of the scaffolding, so it is safe to assume that by then the bridge was complete.

How much the bridge cost is another uncertainty, but contemporary accounts put it at not less than £5,250. Darby appears to have found the additional £2,000 over his original estimate out of his own pocket, and it came close to ruining him. For the rest of his life his property was heavily mortgaged, suggesting that he bought immortality at some sacrifice.

The design of the iron bridge was very conservative and provided a generous margin of safety, for the load-bearing properties of cast iron were not then fully understood. But Darby,

The importance of the Iron Bridge and the surrounding area in the history of industrial development is reflected by UNESCO's decision to declare them a World Heritage Site.

Darby's Metal Masterpiece

or Pritchard, made a small error by failing to allow for the fact that a bridge of iron would be much lighter than one of stone. In a stone bridge, the weight is such that the arch presses outward with great force, requiring embankments to resist movement. In the iron bridge the outward pressure was less, so the embankments moved slowly inward, raising the crown of the arch.

In the early years of the nineteenth century, two additional side arches were added to the bridge, but earth movements continued and cracks were identified. At the beginning of the twentieth century, various additional straps and braces were added to strengthen the structure, which was closed to road traffic in 1934. By the end of the 1960s, continuing movement of the abutments had created some anxiety about the future of the bridge.

Consulting engineers advised that to prevent further movement of the abutments, the foundations of the north abutment should be strengthened and a strut inserted under the water to keep the abutments at a fixed distance apart. The strut took the form of a reinforced concrete slab laid in a trench in the bed of the river, together with walls at each end running up the inside walls of the abutment. Despite flooding and many difficulties, the work was completed during low water in the summers of 1973 and 1974, ensuring the iron bridge will survive for another 200 years.

The successful completion of the bridge in 1779 began a period of bridge-building in cast and wrought iron which lasted a century. By the early 1790s, a huge bridge was built over the Wear at Sunderland, a single arch of 236 feet, yet containing less iron than the bridge at Coalbrookdale. In 1795 floods on the Severn swept away many bridges, but the iron bridge held fast, further increasing its fame and advertising the advantages of cast iron.

The iron bridge itself survived largely because Coalbrookdale, after leading the way into the Industrial Revolution, became something of a backwater. The centres of industry moved to Manchester, Glasgow, Newcastle and other great towns. If traffic over the Severn had continued to grow during the nineteenth century, there is little doubt that the iron bridge would have been replaced by something bigger and more modern; but it did not. As a result, it remains a unique survivor of a bygone age, and the centre of a flourishing museum which has taken over the historic sites of Coalbrookdale.

A bridge "of very curious construction" was how the subscribers described Pritchard's design, based on the principles of carpentry and employing dovetail and mortice-and-tenon joints. No screws or rivets were used.

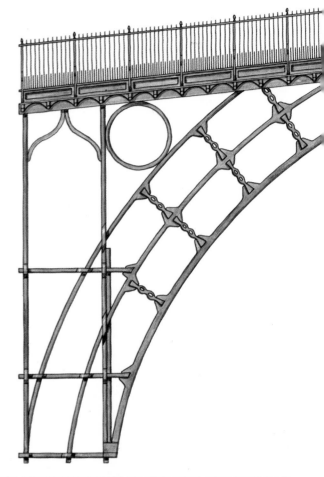

Coalbrookdale and the Iron Bridge were the first industrial sites to become a tourist attraction, visited by many eminent travellers and commemorated in paintings, coins, jugs, tankards, glasses and even fire-grates. Besides the Darby ironworks, the banks of the river and its environs were lined with brick, tile and china works, warehouses and loading wharfs. The bridge was of great value to the district's trade.

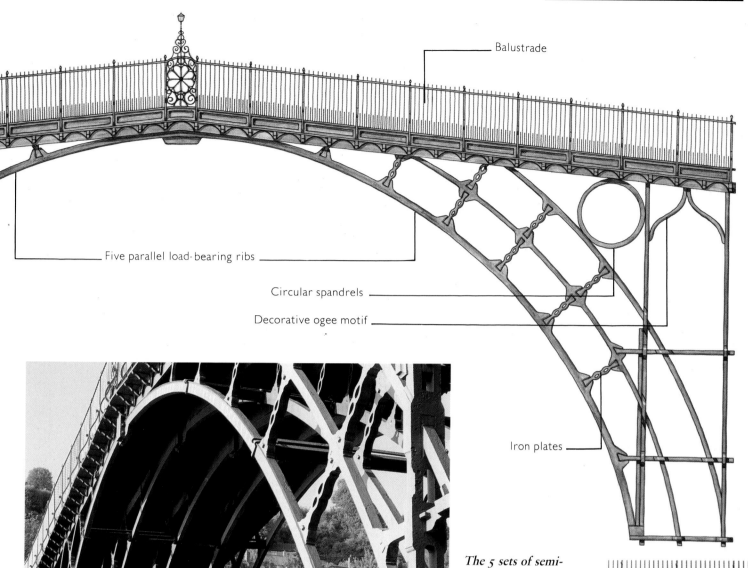

Balustrade

Five parallel load-bearing ribs

Circular spandrels

Decorative ogee motif

Iron plates

The 5 sets of semi-circular ribs *rest against masonry abutments, with 2 further sets of ribs supporting the roadway between the centre and the banks. Joints were secured by wedges—bolts visible today were added later.*

Circles strengthen and decorate the spandrels, *while ornate ogee-headed braces join the 2 vertical members beside the abutments. The design demanded castings of great accuracy.*

The Longest Span

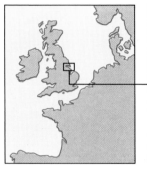

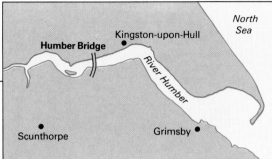

Fact file

The world's longest single-span suspension bridge

Designer: Freeman Fox

Built: 1972–81

Material: Reinforced concrete and steel

Length: 4,625 feet

The world's longest single-span suspension bridge crosses the Humber Estuary, linking together the two halves of the English county of Humberside. It combines the world's longest single span—1,410 metres, or 4,625 feet—with one of the world's fastest accumulating debts. Since the day it opened in 1981, tolls collected from vehicles crossing it have never equalled the interest charges on the money borrowed to build it, so the debt has steadily risen. Critics have called it "the bridge from nowhere to nowhere", and traffic has never come close to meeting the estimates made before it was constructed. It is, none the less, a fine structure, beautiful to look at and a daring piece of engineering.

The bridge was designed by the British company Freeman Fox & Partners, consulting engineers. The main deck is supported by two massive steel cables solidly anchored at each side of the estuary, and carried over the top of concrete towers so that they hang in a graceful catenary curve above the river. The cables are attached by high-tensile steel ropes to a series of shallow boxes which form the roadway. This type of construction is well proven, and is used for bridges requiring very long single spans, such as the Verrazano-Narrows Bridge in New York,

the Bosphorus Bridge in Istanbul, Turkey, and the Severn Bridge linking England and Wales (the last two also the work of Freeman Fox).

The first stage in the construction was to build the anchorages and the towers. On the north bank, both could be built on a solid bed of chalk which comes close to the surface at Hessle. The southern bank was more difficult, with no chalk but a bed of Kimmeridge Clay, 100 feet below the surface, which formed a sufficiently solid base. The southern tower was actually constructed some 510 yards out into the river, with the anchorage at the water's edge. To provide a sound foundation for the tower, concrete caissons were used—huge open circular constructions designed to sink gradually under their own weight as they were built, finishing with their lower edges some 8 yards into the clay. The caissons gradually sank as material was removed from inside them by a grab, but struck underground water which quickly washed away the bentonite, a mineral material used to lubricate the surface of the concrete and ease penetration into the ground. This caused huge difficulties and long delays; ultimately the height of the caissons had to be increased, and 6,000 tons of steel billets temporarily piled on top to create

The Humber Bridge differs from its near rivals in several ways. Topography and geology prevented the usual near symmetry of the side span measurements, though the bridge's length masks the asymmetry. The use of reinforced concrete rather than steel for the towers had been confined to bridges with spans less than half the size.

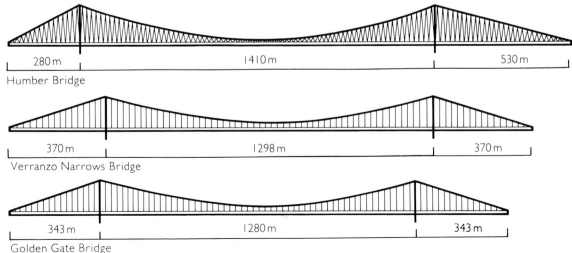

280m | 1410m | 530m
Humber Bridge

370m | 1298m | 370m
Verranzo Narrows Bridge

343m | 1280m | 343m
Golden Gate Bridge

sufficient weight to make the caissons sink to their final positions.

The towers themselves were built of reinforced concrete, poured into forms carried by a platform that could be raised on jacks as the towers went up. Each tower is 158 yards tall, and they rose at a rate of up to 6½ feet a day. The Barton tower, at the southern side of the bridge, was completed in just ten weeks.

The first step in creating the bridge was to throw a footbridge across the river, using six wire ropes carried across in a boat and then pulled into position. The footbridge, a temporary arrangement, was to enable workmen to get into position across the river for the erection of the main support cables which involved some 11,000 tons of wire. During the "spinning" of the cables (as the operation is called, although no actual spinning is involved) the individual strands of wire were carried across in the form of a loop around a wheel carried to and fro by a tramway. Men were stationed at 100-yard intervals along the walkway to handle the wire as it was spooled out. Slowly the cable was built up to its ultimate diameter of 27 inches. At intervals bands were attached around the cables, from which to attach the hangers which would support the decking. The cables were finally coated with red lead paste and wrapped with ⅐-inch (3.5mm) wire, using special wrapping

The length of the span compelled the designers to allow for the curvature of the Earth: the towers are built exactly 36mm out of parallel. Extensive wind tunnel tests were carried out on models of the deck and towers, and provision was made for a total movement and deflection of just under 9 feet between the ends of the spans.

The Longest Span

The anchorages for the cables were constructed of reinforced concrete sections covered by glass-reinforced panels to form a decorative ribbed finish.

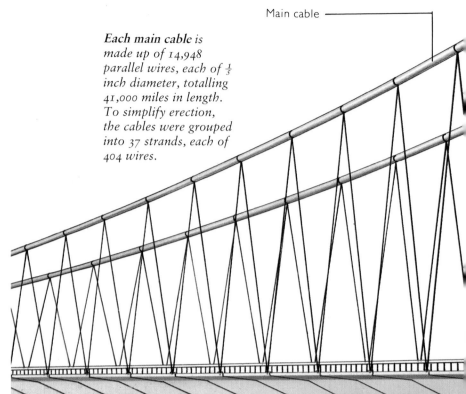

Each main cable is made up of 14,948 parallel wires, each of $\frac{1}{5}$ inch diameter, totalling 41,000 miles in length. To simplify erection, the cables were grouped into 37 strands, each of 404 wires.

machines, before being given five coats of paint to protect them against the weather.

After erection of the bridge's decking, the last stage was to apply the final road surface to the boxes, a $1\frac{1}{2}$-inch (38mm) layer of mastic asphalt, chosen because it is dense enough to stop water penetrating to the steel beneath, and flexible enough to bend to a limited extent without cracking. A total of 3,500 tons of asphalt was needed to complete the roadway.

There is no doubt that the bridge is an engineering and aesthetic success. Although no concessions were made in the design in an attempt to make it beautiful, its austere mathematical simplicity never fails to impress.

The financial statements of the Humber Bridge Board, responsible for running it, are rather less exquisite, consisting as they do largely of red ink. Partly as a result of delays in the foundations for the Barton tower, the bridge was late in opening. Work began on site in April 1973, and the bridge was not opened to traffic until the summer of 1981. The cost also increased enormously, principally as a result of inflation. Instead of the £28 million estimated in 1972, the final cost was £90 million. That, combined with interest during construction, left the bridge with a debt of £151 million on the day it opened. Since then, as interest charges exceed income from tolls, the debt has done nothing but rise. By 1987 the debt was £300 million, by 1989 it totalled £350 million. The Freight Transport Association estimated that if nothing was done, the debt would amount to £576 million by 1993, more than £21,500 million by 2023 and a stupendous £248,247 million by the year 2043. Although tolls bring in around £9 million a year, this is not sufficient, and increasing the tolls would not help as many fewer people would use the bridge. The toll for a car, at £1.60, is already the highest in Britain. However, there is a glimmer of hope that the British government, which lent the money to build the bridge, may agree to write off a proportion of the debt, and give the bridge a chance of breaking even.

The hollow trapezoidal boxes that form the bridge's deck are a distinguishing feature. The steel boxes are lighter than the usual stiffened truss, allowing savings to be made in cables, towers, anchorages and foundations, and the "streamlined" shape reduces wind loading on the bridge. Maintenance is also easier.

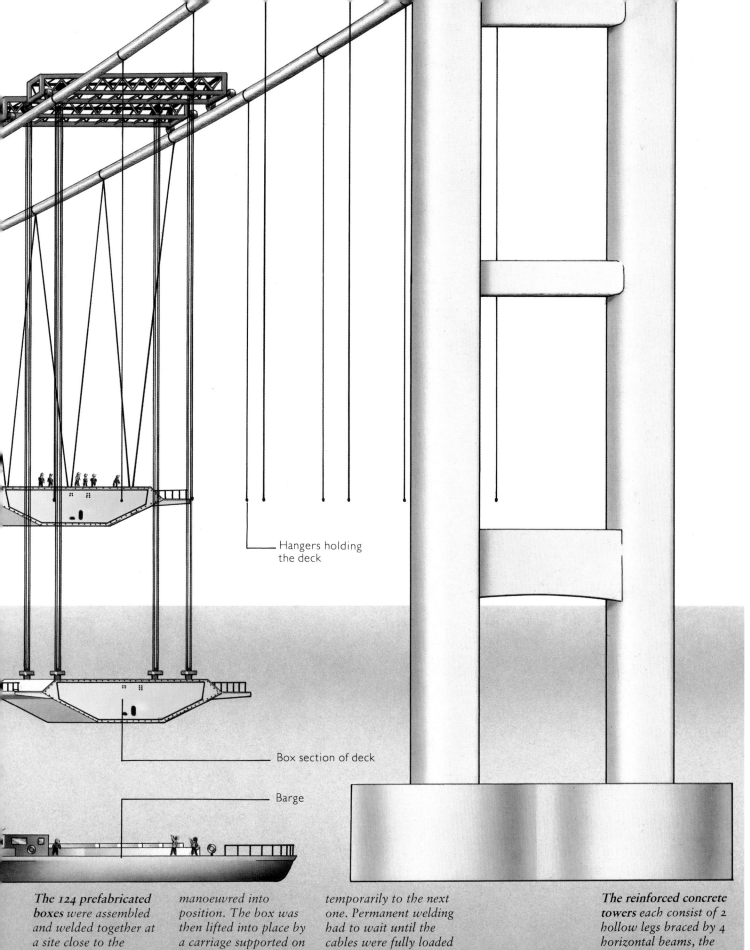

Hangers holding
the deck

Box section of deck

Barge

The 124 prefabricated
boxes were assembled
and welded together at
a site close to the
bridge. They were
floated out one by one
on barges and

manoeuvred into
position. The box was
then lifted into place by
a carriage supported on
the main cables,
attached to the hangers
and connected

temporarily to the next
one. Permanent welding
had to wait until the
cables were fully loaded
and the roadway had
adopted its ultimate
position.

The reinforced concrete
towers each consist of 2
hollow legs braced by 4
horizontal beams, the
lowest one immediately
below the level of
support for the deck.

Bridges of Distinction

Bridges have been built since primitive peoples threw a tree trunk across a stream to produce the first beam, or girder, bridge. The principal distinction between the three main types of bridge—girder, arch or suspension—is the way the forces exerted by the bridge's weight are displaced. In the case of a girder or cantilever bridge (a series of girders balanced on supporting piers), the weight simply rests on the ground. An arch bridge exerts an outward thrust on its abutments and a suspension bridge pulls its cables into tension from the anchorage points sited at each end.

Sometimes the principles are combined, but all bridges are permutations of these basic types. The earliest were built of wood, followed by the use of stone, brick, iron, steel and concrete.

The Sydney Harbour bridge

Built by Dorman, Long of Middlesbrough, England, between 1924 and 1932, the steel arch, supported by granite pylons, was half as long again and required twice as much steel as the largest previously built. The span is 1,650 feet and was built to carry 4 railway lines and a 57-foot wide road. It was tested by 72 locomotives weighing 7,600 tons.

The Great Seto bridge

Opened in 1988 it provides a rail and road link between the largest of Japan's 4 main islands, Honshu, and the smallest, Shikoku. The 6 spans and viaducts total almost 8 miles and it is the world's longest double-decker bridge carrying cars as well as trains. Of the 6 spans, 3 are suspended, 2 are cable-stayed and 1 is a conventional truss. It cost about $8,180 million.

Clapper bridge, Devon

Crossing the East Dart River at Postbridge on Dartmoor, Devon, the bridge was built to carry the Plymouth to Moretonhampstead pack-horse route. It is thought to date from the thirteenth century when the traffic in tin and agricultural produce developed. Built of outcropping moorland stone, huge slabs of unchiselled granite rest on piers and buttresses of the same material. Similar examples are found in Spain, and the oldest surviving datable bridge is of slab construction—the bridge over the River Meles in Izmir, Turkey, which was built about 850 BC.

Luiz I bridge, Oporto

Spanning the River Douro at Oporto in Portugal, the bridge was completed in 1885 to the design of T. Seyrig. He had worked with Gustave Eiffel to produce the very similar Pia Maria bridge close by, opened in 1877. The latter carries a railway line over the single deck whereas the Luiz I bridge bears one roadway on the top of the arch and another provides the tie at the foot of the arch. The Luiz I bridge has a 566-foot arch, and both were built by cantilevering from the river banks. A similar design was used by Eiffel for his railway viaduct at Garabit in France which, at 400 feet above a gorge, was the highest arched railway bridge in the world.

Massive Seabed Structure

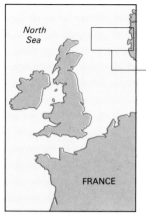

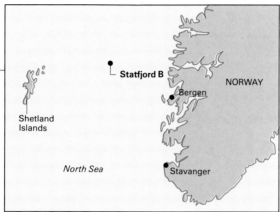

Fact file

When completed, the largest object ever built

Builder: Norwegian Contractors

Built: 1978–81

Material: Concrete and steel

Height: 890 feet

Weight: 824,000 tons

Submerged to the same depth as Statfjord B, the Manhattan skyline would scarcely break above the waves.

In August 1981, the heaviest man-made object ever moved was towed slowly through the western Norwegian fjords toward the North Sea. It was the Statfjord B oil platform, 824,000 tons of concrete and steel, more than 630 feet tall from the oil storage tanks at the bottom to the helicopter deck at the top, and built at a cost of $1,840 million.

Five tugs pulled the huge platform, while three more restrained it from behind to retain control as it wound its way safely through some very narrow fjords. Once in the open sea, the three at the back cast off and the five at the front pulled the platform along at speeds of up to 3 mph. In five days, after a tow of 245 nautical miles, it reached its station, 112 miles due west of Songefjord and 115 miles north east of the

Shetland Islands. Water was pumped into the tanks and the platform settled on the bottom within 50 feet of its planned position.

Statfjord B was then the biggest object produced in an heroic era of offshore engineering. Because the platforms that produce oil from the North Sea are so remote, and so much of them lies below the surface, few people have any conception just how vast they are. From ocean floor to the top of its oil derrick, Statfjord B stands 890 feet, nearly twice as tall as the Great Pyramid at Cheops and not far short of the 1,052 feet of the Eiffel Tower. It is almost 115 times heavier than the latter, nine times heavier than the world's biggest warships (the American Nimitz class aircraft carriers) and three times heavier than each tower in the world's biggest office building, the World Trade Center in New York. Such a structure on land would be an object of huge interest—in the middle of the North Sea, it is almost forgotten.

Statfjord B is a gravity platform, one that rests on the seabed under its own immense weight. The base consists of 24 cells made of reinforced concrete, built in a dry dock in Stavanger. From them rise four hollow legs, also of concrete. Mounted on top is a separate steel structure, the deck, weighing 40,000 tons. This includes all the equipment for drilling the wells and producing 150,000 barrels of oil a day, together with a 200-bed hotel where the workers live, and a helicopter landing pad on the roof. Platforms such as these are production well, refinery, hotel and airport all in one.

The base and the deck were built separately, then mated at sea in an operation which called for supreme precision. The deck, supported by barges, and the base were floated out to Yrkjefjorden, a sheltered deep-water fjord. The top of the four legs in the base matched four short tubes which projected from the bottom of the deck. The trick was to manoeuvre the deck exactly into the right position over the base, then add ballast to the barges to lower the deck while at the same time raising the base by pumping out water from its storage cells. Shifting such huge masses about in the ocean so precisely is a nerve-wracking task. The forces of inertia are so great that the slightest error can chip off huge chunks of concrete as the two huge masses meet. But in 37 hours the entire weight of the deck had been successfully transferred to the base, and the two fastened together by more than 100 4-inch bolts.

Placing the complete platform was another

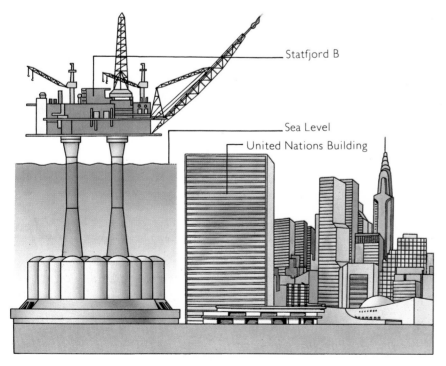

Statfjord B

Sea Level

United Nations Building

Massive Seabed Structure

Statfjord B was the first 4-leg concrete offshore platform to be built. Once the cells of a platform base are completed, 1, 3 or 4 shafts, or legs, are built by the same slipforming techniques employed to build the cells, continuing until the *final height is reached. The shafts have an inner diameter of 75 feet at the base. Two shafts are used for drilling, 1 is the Riser shaft for the oil and the 4th is the Utility shaft, housing loading pumps and ballast water controls.*

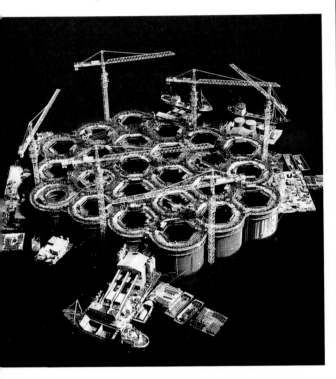

The cell structure at the base contains 24 individual concrete cylinders, or cells, in a concentric formation. Four are extended to form the platform's shafts; the other 20 are storage cells 75 feet wide by 210 feet high for crude oil, which not only assist in smoothing off-loading operations but help to stabilize the platform.

tricky operation. Once manoeuvred into position after its tow, water was pumped into the ballast cells to sink the platform on to the seabed. Around the bottom of the concrete base, a skirt made of steel cut approximately 13 feet into the seabed as the platform settled. Six tugs pulling outward in the form of a star positioned the platform and held it steady while ballast was added, monitored by more than 100 sensors and measuring devices.

Once the skirt began penetrating the seabed, water was pumped from beneath it, and finally the small gaps between the bottom of the base and the seabed were filled by pumping in concrete. The result was a platform placed in the right position and vertical to within a fraction of a degree. It will withstand the worst the North Sea can throw at it, waves 100 feet high and winds of more than 100 mph, without shifting so much as half an inch.

Platforms such as Statfjord B are a world of their own, a universe of noise and power and ceaseless activity. Gas turbines generate enough electricity to run a small town, while inside the huge concrete legs runs a network of pipes and cables of nightmarish complexity. Two legs on Statfjord B are used for drilling the 32 wells, which do not go straight down but curve outwards in a sweeping parabola to reach the farthest corners of the field. Another leg, used for pumps and piping, has 13 separate floors served by lifts.

From the dark oily water of the bilges at the bottom, hundreds of feet below the surface of the sea, you can look upwards, the poet Al Alvarez has written, "as if from the bottom of one of Piranesi's imaginary prisons—a vast enclosed shadowy place, with gangways and galleries and ominous, purposeful machinery, all of it disproportionate to the human scale".

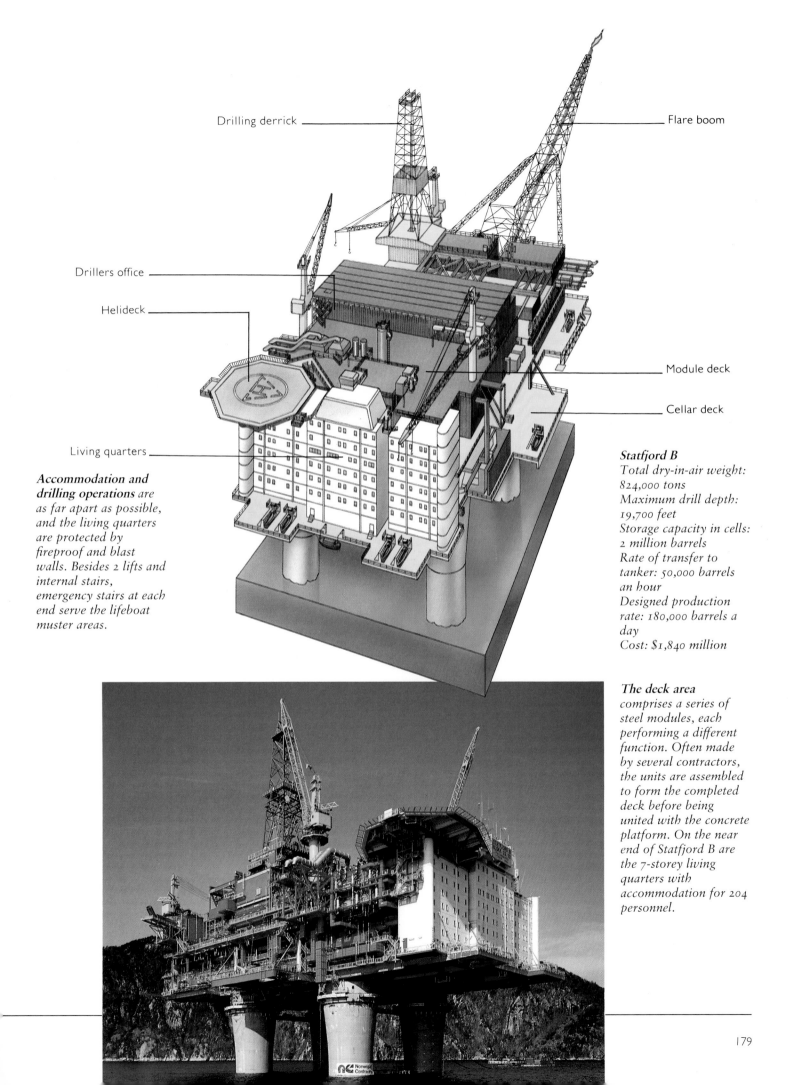

Drilling derrick

Flare boom

Drillers office

Helideck

Module deck

Cellar deck

Living quarters

Accommodation and drilling operations *are as far apart as possible, and the living quarters are protected by fireproof and blast walls. Besides 2 lifts and internal stairs, emergency stairs at each end serve the lifeboat muster areas.*

Statfjord B
Total dry-in-air weight: 824,000 tons
Maximum drill depth: 19,700 feet
Storage capacity in cells: 2 million barrels
Rate of transfer to tanker: 50,000 barrels an hour
Designed production rate: 180,000 barrels a day
Cost: $1,840 million

The deck area *comprises a series of steel modules, each performing a different function. Often made by several contractors, the units are assembled to form the completed deck before being united with the concrete platform. On the near end of Statfjord B are the 7-storey living quarters with accommodation for 204 personnel.*

Generators of the Future

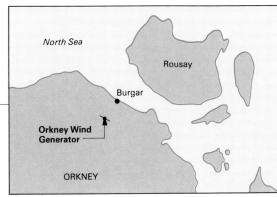

Fact file

The world's most powerful wind generator

Builder: Wind Energy Group

Built: 1985–87

Material: Concrete

Height: 145 feet

Length of rotor: 195 feet

The wind has been used as a source of energy since the seventh century AD, when the first windmills appeared in Persia. From medieval times until the invention of steam power, wind- and watermills represented the summit of technology, machines that could grind corn and pump water, producing far more power than man or animals could. By 1840 there are reckoned to have been about 10,000 windmills at work in England and Wales.

But these beautiful machines have little in common with the new design of wind generators, designed to produce electricity, which have been built since the early 1970s in response to the rising price of fossil fuels and increasing anxiety about the safety of nuclear power. It is highly unlikely that they can ever supply more than about 5–10 percent of the generating needs of developed countries, but even that is a substantial market. At current construction costs, 5 percent of the British power supply would be worth £6,000 million in construction contracts, which explains why a number of big companies have taken an interest.

To make a useful contribution to national electricity generation, as opposed to supplying isolated homes and communities with an intermittent supply, wind generators need to be big. The first attempt to produce such a machine was made by an American engineer, Palmer Putnam, in the 1940s. On the top of a mountain called Grandpa's Knob 2,000 feet up in the Green mountains of central Vermont, Putnam built a tower 110 feet high with a two-bladed, propeller-shaped rotor mounted at the top. It was designed to generate 1.25 megawatts, a substantial output, and went into service on an experimental basis in October 1941.

The generator was not a huge success, suffering many breakdowns and finally failing in March 1945 when one of the blades flew off. The experience at Grandpa's Knob has been repeated many times since, because the fluctuating forces at the root of the blades produce metal fatigue which causes fracture and destruction of the machine. Recent experience suggests that this problem has yet to be entirely overcome.

Where Palmer Putnam went, many other engineers have followed. The world's biggest wind generator was inaugurated in November 1987 on top of Burgar Hill, on the island of Orkney off the coast of Scotland. The generator, LS-1, has a two-bladed rotor mounted at the top of a concrete tower 145 feet high. The machine was designed to generate a peak output of 3 megawatts, and sufficient electricity year-round to supply 2,000 homes connected to the Orkney grid. It cost £12.2 million to build.

LS-1 was constructed by the Wind Energy Group, a joint venture involving Taylor Woodrow, GEC, and British Aerospace. The huge rotor was built by British Aerospace at its Hatfield plant. The tower had meanwhile been constructed using conventional slipforming techniques. On top of it was placed a steel top, or frustum, 20 feet high and weighing 33 tons, which houses the electrical generator. This was made by Seaforth Maritime in Scotland. To it was attached a 66-ton nacelle, built by British Aerospace, which carries the rotor, primary gearbox, bearings, and brake.

The Orkney machine, like all large wind generators, is designed to produce a steady output under a range of wind speeds. In breezes of less than 15 mph, or Force 3 on the Beaufort Scale, it produces nothing. When the wind rises above that speed the rotor cuts in, and reaches maximum power of 3 MW at winds of 37 mph.

At high wind speeds of more than 60 mph, or Force 10, the machine cuts out to avoid damage. The entire construction is designed to survive hurricane force winds of 155 mph. At an average mean wind speed of 24 mph at the hub height of the machine, it will generate 9,000 MW hours per year—equivalent to a continuous 1 MW output 24 hours a day, every day of the year.

Power from the rotor is first transmitted to the primary gearbox inside the nacelle. The output from this gearbox runs downward into the frustum, through a secondary gearbox and into the generator. The rotor turns at a speed of 34 rpm—about as fast as a long-playing record on a turntable—and the two gearboxes multiply the speed up to the generator's 1,500 rpm. The generator speed is fixed by the need to synchron-

Generators of the Future

ize the output with the 50 cycle per second mains supply. The machine, which is intended to be experimental, is extensively equipped with sensors to measure power output, rotor blade loads, wind speeds and other variables. The signals are fed by fibre optic cable to a computer-based data acquisition system located some 100 yards from the machine.

Although LS-1 is the largest wind generator to date, other machines almost its equal have been built in Sweden, Germany, and the US. So far, experience with the big machines has been mixed. The fatigue problem has defeated several of them, and there have also been difficulties synchronizing their output to the grid, which is vital if they are to become reliable suppliers of power. The grid supplies an alternating current which reverses direction 50 times a second, and new power stations have to get into step with all the others when they come on line, so that the peaks and troughs of the supply coincide.

There have been difficulties achieving this with wind-power machines because their speed is not absolutely constant. The force of gravity on the blades makes them slow down slightly when they are moving between 6 o'clock and midnight, and accelerate between midnight and 6 o'clock. This can produce fluctuations in output which make synchronization difficult. In LS-1, a hydraulic coupling between the rotor and the gearbox damps down the oscillations.

Experience with smaller machines has so far been rather better. The greatest concentration of them is in the state of California, where "wind farms" already supply enough energy for 20,000 homes, saving 2.2 million barrels of oil a year. One prime site is the Altamont Pass, near San Francisco, where winds blow almost continuously, and thousands of wind-power machines dot the landscape. By 1988, California had 16,000 wind turbines supplying electricity, most of them in the 150–300 kilowatt range, much smaller than LS-1. Tax incentives have helped the developments in the US, as they have in Denmark and the Netherlands.

In the long run, the future of wind energy will depend on reliability, cost, and environmental acceptability. Most cost estimates at present suggest that wind-generated electricity is roughly competitive with coal and nuclear generation, but certainly not dramatically cheaper. Operating experience suggests that big machines will also need a lot of servicing, spending perhaps a third of their lives out of

commission. Costs of small- to medium-sized machines are, however, falling as more are built, which could improve the economic prospects.

The environmental acceptability of wind power is another key issue. While enthusiasts for alternative energy sources tend to favour wind power over oil, coal or nuclear energy, it is not yet clear that the general public would be happy to see every windy site, and many seashores, occupied by huge wind turbines. To make inroads into total energy consumption, hundreds or even thousands of machines like LS-1 would have to be built, each requiring a substantial area. Such machines cannot be built right next to each other because that reduces the wind and increases buffeting. Wind farms may be green, but is their effect on the environment entirely benign? That is an intriguing and difficult question to which environmentalists may yet have to find an answer.

LS-1 under construction on Burgar Hill (above). Orkney is known for its bitter weather and regular gales, making the site ideal. Though not the first wind generator on Burgar Hill, it is much the largest. The forerunner of LS-1 can be seen beyond the construction site; the MS-1 has an output of 250kW compared with LS-1's 3MW.

The rotor is built in 5 sections, of steel and glass fibre-reinforced plastic and weighs 63 tons. Measuring 195 feet from tip to tip, it is just longer than the wing span of a Boeing 747 jumbo jet. The enormous forces on the rotor make it the most vulnerable component to metal fatigue.

Assembly of the rotor took place in a large hangar where it was tested both statically and dynamically before being taken apart and shipped to Orkney. The angle or pitch of the rotor can be adjusted by hydraulic power (below), the mechanism fitting into the hub. The outer 30 percent of the blades can be feathered to adjust their speed for a constant output.

An internal hoist enabled the rotor to be lifted up to be united with the primary gearbox in the nacelle. The 65-ton nacelle and 32-ton frustum, containing the generator, had to be put in place by crane. A 4-person elevator provides access to the nacelle for operations and maintenance personnel.

Nuclear Goliath

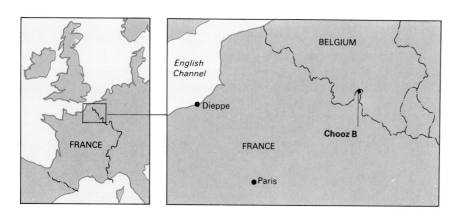

Fact file

Europe's largest nuclear power station

Builder: Electricité de France

Built: 1982–

Output: 2,800 megawatts

Area of site: 331 acres

No other form of engineering combines power and precision in quite the same way as do nuclear power stations. They are vast machines, costing hundreds of millions of pounds, and powerful enough to supply all the electricity for a huge city. Yet they are put together with a clockmaker's precision, and in conditions of cleanliness that would do credit to an operating theatre. The biggest in Europe, and among the very biggest in the world, is Chooz B, which, when finished in 1993, will generate 2,800 megawatts from two pressurized water reactors set in a curve of the River Meuse in France close to the border with Belgium.

The years since 1979 have not been happy ones for nuclear power. The accidents at Three Mile Island in 1979 and Chernobyl in 1986 have brought home the consequences of error in a technology that offers much, but at a price. Many nations have stopped building nuclear plants altogether. Not so France, which possesses uranium but lacks indigenous sources of coal or oil and embarked in the early 1970s on an ambitious plan of nuclear investment.

In 1973 France produced less than a quarter of its energy from its own resources. By 1986, it produced 46 percent, and it expects to exceed 50 percent during the 1990s. This has been achieved by building nuclear plants on all the main rivers of France and along the coastline. Today France has more than 50 reactors in operation. The three biggest, still under construction, are at Chooz on the Meuse and at Civaux on the Vienne. At Chooz, where there is already a much smaller reactor, Chooz A, there will be two 1,400 MW reactors; at Civaux, one.

Part of the French success in this huge and costly programme has come from building a single type of reactor, increasing in size only gradually, and gaining enormous expertise in putting the reactors together. The pressurized water reactor was originally developed by Westinghouse in the United States, based on a scaled-up version of the reactor used to power nuclear submarines. But the French had little hesitation in abandoning their own home-grown designs in favour of a reactor made in America. The design has since been considerably developed.

All nuclear power stations have certain things in common: uranium fuel, usually in the form of pellets of uranium dioxide; a coolant to remove the heat produced by the nuclear reaction, and generate steam; and a moderator, the purpose of which is to slow down the neutrons produced by nuclear fission and improve the functioning of the reactor. In a PWR, ordinary water is used as both coolant and moderator. The heat generated by the fuel is transferred to the water inside a steel vessel at a pressure of about 2,000 lb per square inch, or 130 times normal atmospheric pressure. The water temperature rises to more than 570°F (300°C), but the water does not boil because it is under such great pressure. Instead it flows in tubes through a steam generator, where it gives up its heat to a second set of tubes, also carrying water. This water, which is not pressurized, does boil, producing steam which is then fed to turbo-generators to produce electricity.

The most critical component in a PWR reactor is the pressure vessel, for if it should fail catastrophically there would be a massive leak of radioactive material into the reactor building, and perhaps also to the outside world. The pressure vessels at the Chooz B plant are typical. They are cylinders 44 feet high and with an inner diameter of 14 feet 6 inches, made from steel almost 9 inches thick, and weigh 462 tons. On top is a dome-shaped lid which is bolted firmly down during normal operation, but which can be removed when fuel needs replacing—once every year at Chooz B. The standards of manufacture of the pressure vessel and pipework

Nuclear Goliath

which carries the cooling water in and out need to be of the very highest. So far, the care has paid off: there have been no catastrophic failures of pressure vessels in commercial nuclear reactors.

Inside the pressure vessel is a network of fuel rods, each about 14 feet long and less than $\frac{1}{2}$ inch in diameter. Inside each rod are pellets of uranium oxide in which the proportion of fissile uranium-235 has been artificially increased to about 3 percent. The rods are arranged into fuel assemblies, which each contain 264 rods, held together on a grid with spacers. The Chooz B reactor has 205 such assemblies—a total of 54,120 fuel rods. The fuel assemblies occupy the lower half of the pressure vessel and are surrounded by a cylindrical sheath. Water flowing through pipes at the top is directed down to the bottom of the pressure vessel between the sheath and the walls, and then flows upward through the gaps between the fuel rods in the centre. As it does so it removes heat from the fuel rods and increases in temperature before flowing out through a separate set of pipes to the steam generator.

The PWR is a very compact design, with all the heat generated in a comparatively small volume. It is therefore vital to maintain a flow of water at all times, because if it were to fail the reactor could overheat and melt down. Coolant flow is necessary even after the reactor has been shut down, for it still continues to generate intense heat through radioactive decay. This is why PWRs are always fitted with emergency core-cooling systems, independent of the regular circuits, to ensure adequate cooling at all times.

The much bigger structures that tower over the reactor buildings at Chooz B are cooling towers. Due to the laws of thermodynamics, not all the heat generated by the nuclear fuel can be turned into electricity – more than half of it must be discarded.

The purpose of cooling towers is to remove the heat from the water and disperse it into the air. Air flows in at the bottom and rises naturally to the top as a result of the chimney effect of the structure. Meanwhile, warm water is sprayed over a network of vanes between which the rising air flows. The result is to heat the air— which leaves from the top of the tower as a plume loaded with water vapour—and to cool the water. The cooled water is then released into the River Meuse; but even now it is not quite cold, and will raise the temperature of the river by 1.8°F (1°C). This, claims Electricité de France,

is too little to have any measurable effect on the life of the river.

The final key element in the plant is the turbo-generator which turns the steam into electricity. At Chooz B, the turbo-generators are among the biggest ever built, capable of generating 1,400 MW and weighing 3,150 tons. At one end is a steam turbine, in which steam is allowed to expand through series of fans arranged along a common shaft, forcing it to turn at 1,500 rpm. Attached to the same shaft at the other end of the machine is an electrical generator which produces the electricity.

Chooz B will take ten years to complete, and is expected to cost 15,000 million francs ($2,640 million) at 1985 prices. Building it will employ 1,600 people; operating it will need between 500 and 550. Work began on site in July 1982 and the first reactor is expected to come on stream in 1991, with the second following in 1993.

Cooling towers are a feature common to coal-fired and inland nuclear power stations. Nuclear plants by the sea generally dispose of waste heat by allowing it to flow away in the form of hot water into the ocean, where its effect is minimal. River-based plants cannot do this, because the river water would soon be lukewarm and cause enormous ecological damage.

The foundation ring of a cooling tower (below), showing the anti-turbulence, finned pillars, which look too insubstantial to support the concrete tower. Heat exchanging machinery covers the floor of the tower.

The reactors are surrounded by a concrete vault housed in a double-walled structure (above), the containment building. This provides successive layers of protection to isolate radiation from the outside world in the event of a leak.

Part of a steam generator (right) which utilizes heat generated by the PWR to produce steam in a secondary system, which in turn powers the turbo-generator. Each generator contains 5,600 U-shaped tubes.

The dome over the PWR has an inner and outer casing, strengthened by anti-fracture steel reinforcement (above). Almost 261,500 cubic yards of concrete are required for each reactor and fuel building and its associated cooling tower.

Harness of the Sun

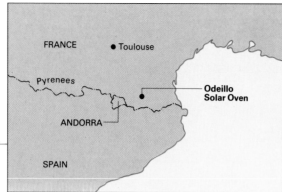

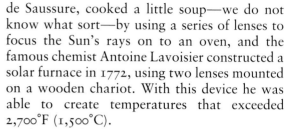

Fact·file

The largest solar-powered generating plant in Europe

Builder: National Centre for Scientific Research

Commissioned: 1969

Material: Glass

Area of mirror: 20,000 square feet

The tower in front of the fixed mirror houses the furnace which receives the concentrated solar rays from the mirror.

The energy that reaches the Earth from its star, the Sun, is many times greater than that actually used by the humans on our planet. The amount that falls just on the roads and freeways of the United States is more than double the world's total consumption of coal and oil. But solar energy is widely spread: at best, the amount falling on an area of 1 square yard on a sunny day is around 1,000 watts. If the weather is cloudy, it might be only a fifth of that.

To make effective use of the Sun's energy, it has to be concentrated. This principle has been understood since ancient times. Archimedes is supposed to have attacked the Roman fleet at Syracuse in 214 BC by reflecting solar rays on to their ships with a system of mirrors arranged on the shore. The Athenians and the Aztecs lit their sacred flames with the help of concave mirrors, and countless campers have done the same for their bonfires with convex lenses. In the eighteenth century a Swiss physicist, Horace-Benedict

de Saussure, cooked a little soup—we do not know what sort—by using a series of lenses to focus the Sun's rays on to an oven, and the famous chemist Antoine Lavoisier constructed a solar furnace in 1772, using two lenses mounted on a wooden chariot. With this device he was able to create temperatures that exceeded 2,700°F (1,500°C).

The use of solar energy in this way seems to have been particularly fascinating to the French. In 1945 the chemist Felix Trombe, who specialized in the study of high-melting refractory materials, was asked by the French National Centre for Scientific Research (CNRS) to study the subject in depth. He used an old radar dish 6 feet in diameter, silvered on the inside to reflect light, to achieve temperatures of more than 5,400°F (3,000°C). It was as a result of his experiments that a laboratory was set up in the western Pyrenees, at the fort of Montlouis, to continue the work. Close by, at Odeillo, was built one of the most impressive and successful of solar furnaces. It came into service in 1969.

The Odeillo furnace uses a series of flat mirrors, arranged on terraces on a hillside, to reflect the Sun's rays on to a parabolic mirror 138 feet wide. There are 63 flat mirrors, known as heliostats, arranged on eight terraces. The heliostats can turn both horizontally and vertically to track the Sun and ensure that its rays are always reflected exactly on to the central parabolic mirror. The movement is carried out automatically, under hydraulic power. A computer controls the hydraulic system.

The parabolic mirror is fixed, and focuses the light falling on to it. It is a characteristic of parabolic mirrors that parallel rays of light reflected from them all pass through the same point, called the focal distance; at Odeillo this point is 58 feet in front of the parabolic mirror. The mirror itself consists of 9,500 small mirrors, each just under 18 inches square, creating a total surface area of some 20,000 square feet. The light from this mirror is focused on to the furnace, which is mounted in a tower fixed at the focal point of the mirror. All the heat is not concentrated at a single point, but into an area just 16 inches across. This produces immensely high temperatures, up to 6,870°F (3,800°C).

Because the air in the Pyrenees is clean, little dirt is deposited on the mirrors and any that is tends to be removed by frost and snow. The heliostats need cleaning only every two years or so, while the parabolic mirror was cleaned only

Harness of the Sun

twice in its first 16 years of operation. The furnace operates for around 1,200 hours every year and requires little servicing of its hydraulic and electronic control gear.

The advantage of the solar furnace is the intensity of the heat source and its great purity. Unlike other methods of reaching such high temperatures, there is no danger of contamination, which has made it useful for the production of materials such as a vanadium oxide semiconductor used by Kodak in certain photographic films. It can also be used to investigate the resistance of materials to sudden changes of temperature. By using specially cooled shutters, the heat can be turned on and off in a period as short as $\frac{1}{10}$ of a second, creating very rapid cycles of heating and cooling which would cause most materials to shatter. The thermal tiles used to protect the American space shuttle when it re-enters the atmosphere, and the protective material on missiles have been tested in this way.

There is, of course, another way in which the concentrated heat of the Sun can be used. By raising steam, it can generate electricity and replace coal, oil or nuclear generating plants. Near Odeillo there is a 2.5 megawatt plant based on this principle, using 200 mirrors arranged in a semicircle to reflect sunlight towards a tower 328 feet high. This is the largest solar-powered generating plant in Europe, but there is a much bigger establishment at Barstow in California, named Solar 1.

This plant uses 1,818 mirrors to focus the sunlight on to a boiler at the top of a tower 255 feet high. First tested in 1982, Solar 1 is capable of generating 10 megawatts, and cost $141 million to build. It is situated in the Mohave Desert, which enjoys more than 300 cloudless days a year. But it occupies about 100 acres of land, which is an indication of just how much space is needed for a solar power station, even in ideal circumstances. To generate the whole of the United States' power supply in this way would mean occupying much of the desert lands with power stations. Nations less favoured with solar energy, like Britain, could simply not find the space to justify such plants.

Unlike Odeillo, the Barstow plant does not have a complex parabolic mirror. Instead, its sun-tracking mirrors focus the light on to a single receiver mounted on top of the tower. The receiver consists of a series of pipes, painted black to absorb energy better, through which flows a fluid. In the simplest plants the fluid may

be water, which is converted to steam by the heat, and the steam used to generate electricity in conventional turbines. An alternative is to use molten salts, which are an excellent medium for storing heat, and divert the problem of handling high-pressure steam up at the top of the tower. The salt is heated by the concentrated solar rays, and flows through pipes to the ground, where it gives up its heat in heat exchangers to water, generating steam to produce electricity.

The Barstow plant has proved very successful, and has been followed by a series of schemes in the US to replace the conventional boilers of power stations with "power towers" using solar energy. One survey in the 16 south-western states suggested that up to 13,000 megawatts of solar energy could be used in so-called repowering schemes, replacing about 11 percent of the oil and gas being consumed at present by the electricity utilities.

The size of the mirrors *necessary to generate a commercially worthwhile amount of power (above), and their obtrusiveness in the landscape, militate against their adoption on a large scale. However, as the costs of fossil fuels rise, harnessing the Sun's energy will become an increasingly attractive prospect.*

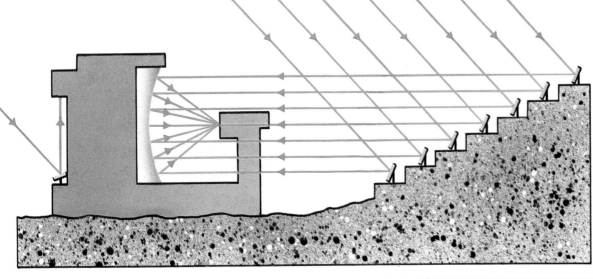

The intensity of the solar radiation at the focal point of the mirror's rays is 12,000 times more powerful than normal solar rays when the furnace is operating at maximum capacity; in average conditions it is 2,000 times more powerful.

Banks of heliostats (right)—flat mirrors that reflect the Sun's rays on to the main, parabolic mirror—require a large amount of space, and need to be in a stepped pattern to maximize their effectiveness. At Odeillo, the heliostats are each 24 feet by 20 feet, giving a total area of collection of 30,000 square feet.

The number of overcast days in temperate climates (left) is the principal drawback to the exploitation of solar energy in such climates. Yet the Sun's energy is so great that its potential compels extensive research: in 2 weeks, the solar energy falling on Earth is the equivalent of the world's entire initial reserves of coal, oil and gas.

CHAPTER THREE

Underground Engineering Achievements

e know more about the surface of the Moon than we do about the ground 10 miles beneath our feet. Despite every achievement of engineering, the Earth's crust remains unknown territory into which we have scarcely ventured. The mantle and the core that lie below the crust are even more remote. The deepest wells drilled go down no more than 10 miles, and then only with huge difficulty. Attempts to reach the Mohorovicic Discontinuity—the point at which crust gives way to mantle—were abandoned in the 1960s when the costs of the venture soared out of control.

It was an experience familiar indeed to tunnel engineers. As the barrier that has divided Britain from the rest of Europe for so long is finally undermined by the Channel Tunnel, the costs and complexities of boring tunnels are constantly rehearsed. There is hardly a tunnel that has ever been dug, anywhere, without exceeding its budget. Even in Japan, where the world's longest rail tunnel was completed during the 1980s at huge and uncovenanted expense, construction had taken so long that the airlines had taken most of the traffic. Moreover, the cost overrun meant that there was no money left to build the railway for "bullet" trains that might have won the passengers back. No one, it seems, can produce a formula that is entirely right.

The sinister nature of the subterranean world, and the sense of fear it provokes, are nowhere stronger than in the catacombs, where early Christians were laid to rest like lost souls each to its own filing cabinet. A similar unease pervades the long tunnels of the world's greatest underground factory, in the Harz Mountains of Germany, where slave labourers from the conquered nations were put to work building a new and terrible weapon, the ballistic missile, during World War II. Compared with these, the light and cheerful tunnel that houses the world's biggest atom-smasher at CERN, the European particle physics laboratory, seems wholly beneficent. Here a vast machine buried in the stable rock is used to probe the fundamental nature of the universe. Tiny particles too small to imagine whirl round at fantastic speeds before smashing into one another, a concept as peculiar in its way as the ancient myths that once gave the nether world such notoriety.

Underground Engineering Achievements
Catacombs
Nordhausen V2 Factory
LEP Accelerator
Seikan Rail Tunnel
Great tunnels

Subterranean Mausoleum

Fact file

The early burial chambers of Christians provide one of the world's most extensive networks of underground passages

Built: 2nd–5th centuries

Number of sets: 42

ITALY

Rome

The Catacombs

On the outskirts of Rome lies a honeycomb of subterranean passages where the bodies of the early Christians were buried. The catacombs, as they are called, date from a period when being a Christian was a dangerous activity. Here were laid the bodies of some of the early popes, and of Christians martyred by Roman emperors determined to stamp out the Church. Alongside them are ordinary Christians, men and women who shared the conviction that burial of the dead was a supreme duty if they were to participate in the resurrection in which all Christians believe.

The digging of the catacombs, a process which went on for several hundred years, was the task of a group of men called *fossores*, or diggers. The marks of the picks they used to carve out the passages in the soft rock can still be seen today. By the middle of the third century, when the Church was under very strong pressure, many men must have been employed as *fossores* in order to have created such a labyrinth of tunnels. About 40 different sets of catacombs are known, most of them close to the main roads that run into the city. The total length of tunnel is difficult to calculate, since the catacombs run to and fro like a maze, usually occupying several levels, but they amount to many miles.

The *fossores* who created them lived a cold and gloomy life, hacking away in narrow tunnels with only the dead for company; it was no task for the faint-hearted. Sometimes they would be required to carve out underground rooms, 10 feet or more square, which served as crypts for an entire family. There seems little doubt that some of them supplemented their incomes by stealing anything of value from older graves.

Later, as waves of invaders swept through Rome, the very existence of the catacombs was forgotten, and they remained unvisited for hundreds of years. They were then rediscovered at the beginning of the seventeenth century by an enthusiast named Antonio Bosio, who appears to have spent most of his life after the age of 20 searching for them. He would set out on foot from the centre of Rome and spend whole days looking for entrances to the catacombs. He rediscovered about 30 of them, and published his results in his book *Roma Sotterranea*, Subterranean Rome. Proper archeological study followed in the nineteenth century. In 1854, when Pope Pius IX was told by the archeologist G.B. de Rossi that the graves of the early popes had been rediscovered, he initially refused to believe it. Inscriptions, however, left no doubt that these were, indeed, the last resting places of five popes from the third century.

Why did the early Christians go to such enormous trouble to bury their dead? Catacombs are not, in fact, an exclusively Christian custom, and they are found all over the Mediterranean, particularly in Malta, Sicily, Egypt, Tunisia and Lebanon. But the fact that after his crucifixion Christ's body was laid in a sepulchre, with a stone to cover the entrance, must have helped make the idea popular to his followers.

Another reason was undoubtedly the danger of persecution. Under the Emperor Valerian, for example, Christians were forbidden to visit cemeteries, and all burials within the walls of Rome were forbidden by law. To protect the mortal remains of the Christian martyrs, they were taken from ordinary graves and concealed in catacombs where they were less likely to be disturbed. These burial places then became centres of pilgrimage, and ordinary Christians began to express the wish that when they died they should be laid as close as possible to the bodies of the martyrs.

A further consideration may have been limitation of space, encouraging a system of burial where successive tombs are buried ever deeper in the ground. A final factor was the wish of living Christians to visit the dead on the anniversary of burial, there to celebrate the Eucharist. For a persecuted Church it was clearly much easier to do this privately underground than in a more conventional cemetery. There is, however, no reason to suppose that the catacombs were used as secret places for worship. The largest rooms in any of the catacombs can accommodate no more than 40 people, and by the third century there were at least 50,000 practising Christians in Rome, so an underground service would not have been very convenient.

The ground around Rome was ideal for tunnelling, consisting of a soft *tufa* that was

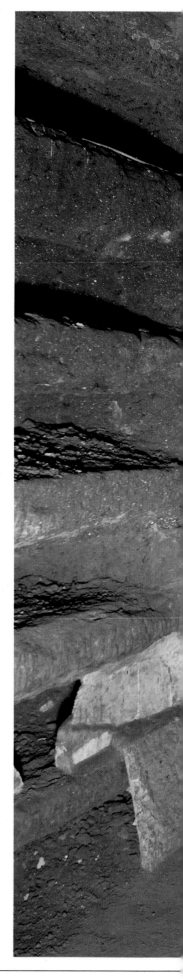

Subterranean Mausoleum

The Catacombs of St Calixtus (right) were the official burial place of the bishops of Rome, and named after Callistus who was put in charge of the cemetery by Pope Zephyrinus. Callistus was himself elected *pope after 18 years administering the cemetery. The catacombs are on five levels and contain many frescoes. The papal crypt still has Greek inscriptions of the early martyred popes from the 3rd to 4th centuries.*

The Columbarium in the privately owned Vigna Codini (above) has space in its largest room for 500 urns containing cremated remains. The term "Columbarium" was applied to such rooms on account of their resemblance to the holes in the walls of a dovecote, which is the principal meaning of the word, "columba" being the Latin for dove.

quarried to form the basis of a strong mortar used by the Romans for building. Often a catacomb began at an entrance into a hillside originally dug by quarrymen to extract *tufa*. A passage was dug into the ground, and new passages cut at right angles to it, perhaps leading to further passages parallel to the first one. Sometimes the passages, 7–10 feet high and 3–4 feet wide, went down for as many as five distinct levels, with light wells drilled to the surface to let some sunshine penetrate the gloom. Neighbouring catacombs often joined up, creating a veritable warren of passages, in which it is easy to get lost.

The simplest grave was a kind of shelf, dug into the walls of the passage, in which the body was laid, wrapped in two layers of linen. The space was then closed by sealing it with tiles. Such graves are known as *loculi*. Alternatively, there were tomb chambers, or *cubicula*, in which

a whole family might be buried, the equivalent underground of the family vaults in an ordinary cemetery. Along the passages oil lamps were cemented into the walls, together with vases for perfume to sweeten the air. Sometimes a child's toy, or a coin, was also cemented into the wall close to a tomb. Most of the tiles have been torn off by grave robbers.

What little we know about the *fossore* who dug these catacombs comes from the drawings that are found on the walls. They wore a short tunic, and carried a lamp with a chain and spike attached, so that it could be driven into the wall and hung to supply light. They also carried a basket for removing the material they hacked from the walls with their picks. *Fossore* clearly regarded themselves as more than mere gravediggers, with a status approaching that of the clergy. They were also artists, decorating the catacombs with simple carvings and paintings of

A catacomb near Via Latina (left) was discovered as recently as 1955 during building work. It is thought to date from between AD 320 and 360 and to have been the preserve of a few wealthy families. Some frescoes depict subjects never before encountered in the catacombs, such as Greek mythology. This picture in a cubiculum depicts Alcestis after being rescued from Hades by Hercules who returns her to her husband Admetus.

no great merit, but fascinating for the light they cast on the early Church.

In the period when Rome was attacked by successive waves of invaders, the bodies of many of the Christian martyrs were removed from the catacombs and taken into the churches and basilicas. In AD 609, for example, 28 wagon loads of relics are said to have been taken to the church of S. Maria ad martyres, and the Lombard invasion of 756 did enormous damage to the cemeteries outside Rome, encouraging more relics to be brought into churches in the city by Pope Paul I (757–67).

Paschal I (817–24) had the bodies of 2,300 martyrs transferred to the church of S. Prassede. By this time, the catacombs were almost empty, and began to be forgotten. From the ninth century, until the time of Bosio in the seventeenth, they slumbered quietly, visited by very few. Bosio's work encouraged others to venture down, but few were willing to believe that these were really Christian burial places. Many relics and paintings were destroyed by enthusiastic but amateurish exploration.

The paintings that do survive in the catacombs are important because they are almost the only remaining forms of Christian art from the era when the Church was persecuted. Early church buildings have all perished, but the simple decoration of the catacombs survived. Many of the paintings depict scenes from the Old Testament—the Fall, Noah's Ark, and Abraham's sacrifice of Isaac are particularly common.

There are also many New Testament scenes, such as the baptism of Jesus, and many of his miracles. By far the most common of these is the raising of Lazarus, of which more than 50 examples have been found in the catacombs—an understandable theme to find in a place devoted to the dead.

The Catacombs of Saints Pietro and Marcellino (above) are the most pictorially decorated in Rome. Representations of the Fall are common, not only in paintings but on sarcophagi and on glass cups. A version in the Via Latina catacombs shows Adam and Eve being expelled from Eden by God, who was portrayed without inhibitions in early Christian art.

Factory beneath a Mountain

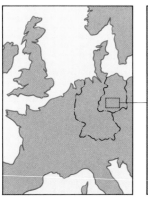

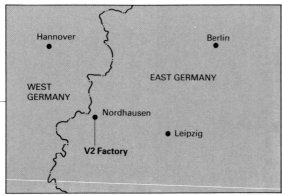

Hannover

Berlin

EAST GERMANY

WEST GERMANY

Nordhausen

Leipzig

V2 Factory

Fact file

The world's largest underground factory

Builder: Third Reich

Built: 1936–42

Length of tunnels: 7 miles

Area: 1.27 million square feet

On the evening of 8 September 1944, without a hint of warning, a huge explosion blew a hole 20 feet deep in the middle of Staveley Road in the London borough of Brentford and Chiswick. Three people were killed, one a young soldier walking down the road, another a three-year-old baby. They were the first-ever victims of a ballistic missile, the V-2. It had been launched from a mobile platform parked on a suburban street in The Hague, then under German occupation. Sixteen seconds after the first explosion, a second V-2 landed in Epping, without killing or injuring anybody.

The V-2 was the secret weapon that Hitler hoped would bring him victory in the war. Together with the flying-bomb, the V-1, it was assembled in an extraordinary underground factory in the Harz Mountains, in Lower Saxony. Protected by 200 feet of rock, and with their entrances carefully concealed from the air, the underground tunnels at Nordhausen remained intact and invulnerable until the end of the war, when they were overrun by troops of the US 1st Army. By then they had produced the vast majority of the 1,403 V-2 missiles fired at London, together with many more used against targets in Belgium. The Nordhausen plant is the largest underground factory ever created.

Work began in 1936 when, as part of the German preparations for war, the state-controlled oil storage firm Wirtchaftliche Forschungs GmbH drew up plans for an underground storage depot under Kohnstein Hill near Nordhausen. The rock was anhydrite (calcium sulphate), ideal for the purpose since it is dry, soft, and easy to tunnel, yet strong enough for long, continuous galleries without extra support. The material excavated was used as a raw material for the production of cement, sulphur, and sulphuric acid.

By the time the complex was complete in 1942, there were two service tunnels, more or less parallel to each other and about 180 yards apart, running into the hill. Each tunnel was just over 1 mile long, 35 feet wide and 25 feet high. At intervals of about 40 yards the tunnels were linked by 43 cross-galleries, arranged like the rungs of a ladder, each about 30 feet wide and 22 feet high. The 7 miles of tunnel gave a floor area of 1.27 million square feet. Huge cylindrical tanks were installed in the cross-galleries for storing fuel, but in 1943 these were removed on the orders of the German Ministry of War Production, and the whole complex requisitioned for the production of weapons.

After successful Allied air raids on Hamburg, the ball-bearing factories at Schweinfurt and the research centre at Peenemünde where the V-weapons were developed, a bomb-proof factory was needed, and Nordhausen was the ideal candidate. The southern half of the complex was handed over to Mittelwerk GmbH for the manufacture and assembly of V-1 flying bombs (less wings) and V-2 rockets (without warheads). The northern part was given to the Junkers company for the assembly of Jumo 004 jet engines for Messerschmitt 262 fighters, and Jumo 213 piston engines for the older Focke Wulf 190 fighters.

Few changes were needed. An electricity supply was laid on from a nearby power station, and a cavern 75 feet high was hollowed out so that completed V-2 rockets could be stood vertically for testing the electrical components. During August and September 1943, prisoners were moved from other concentration camps to provide the work force.

Toward the end of October, the whole camp was moved underground and the prisoners—mostly French, Russian and Polish, but with some German political prisoners, too—were billeted in three chambers which were dark,

damp and full of dust. They slept in bunks four tiers high, working 12-hour shifts. When one shift went off to work, the other tried to sleep in the same dirty bunks, covering themselves with the same blankets. There were no latrines—empty carbide barrels, cut in half, had to serve—and it was a walk of over ½ mile to a water tap.

Albert Speer, the German armaments minister, visited the plant in December and recorded his impressions in his autobiography, published after the war. "The conditions for these prisoners were in fact barbarous, and a sense of profound involvement and personal guilt seizes me whenever I think of them. As I learned from overseers after the inspection was over, the sanitary conditions were inadequate, disease rampant; the prisoners were quartered right there in the damp caves, and as a result the mortality . . . was extraordinarily high."

As a result of Speer's orders, a concentration camp was built above ground to house the prisoners, and conditions improved. More and more prisoners were sent to the camp, until the pool of slave labour reached about 20,000. The SS issued strict orders forbidding private contact between the convicts and the German staff. On

A partially completed V-2 being examined by a US soldier after the capture of Nordhausen by General Omar Bradley's army on 11 April 1945. Hitler had ordered destruction of the entire plant before capture, but it was left almost untouched after the evacuation of the specialists and workers.

Factory beneath a Mountain

no account was news of what was happening at Nordhausen to reach the outside world.

The first three V-2 missiles were delivered from Nordhausen on New Year's Day 1944, and by the end of January another 17 had been completed. From then on production built up rapidly, with 250 missiles being delivered in June. The production of the V-1 began later, in July 1944, and 300 were delivered that month. While the V-2 was a highly complex and sophisticated weapon, the V-1 was cheap and crude. Both were effective, and in London there was a good deal of Blitz humour over which was more terrifying—the V-1, whose engine could be heard until it cut out, leaving an agonizing wait for the bang, or the V-2, which arrived without any warning at all. One London woman, from Streatham, was philosophical about the V-2, regarding it as one might a thunderbolt: "By the time it had arrived, you were either dead or it had missed", she said. But the writer James Lees-Milne, who was living in Chelsea, was in no doubt that the V-2 was far more alarming than the V-1, because it gave no warning: "One finds oneself waiting for it and jumps out of one's skin at the slightest bang or unexpected noise, like a car backfire or even a door slam."

The entrances and ventilation shafts of the factory had all been well camouflaged. The missiles were loaded on to rail wagons or lorries in the tunnels, and canvas covers tied in place. The trains then ran out of the tunnels and on to the German railway network, on their way to the launching pads close to the Channel. These precautions successfully concealed the plant from reconnaissance aircraft, and the only hint of its importance came from interrogation of a German prisoner of war in the late summer of 1944. An American plan to attack the plant by dropping huge amounts of napalm on the tunnels and ventilation shafts, to create a fire that would suffocate those inside, was rejected, fortunately for Nordhausen's slave workers.

During December 1944, a total of 1,500 V-1s and 850 V-2s were produced at the underground plant, and its success and invulnerability led to proposals to increase its floor area sixfold. New tunnels were begun to house a plant to manufacture liquid oxygen (one of the fuels used by the V-2), a second factory for aero engines, and a refinery to produce synthetic oil. But all came to an end on 11 April 1945, when US forces reached the area. They stayed for about six weeks, carrying out a minute examination of the factory

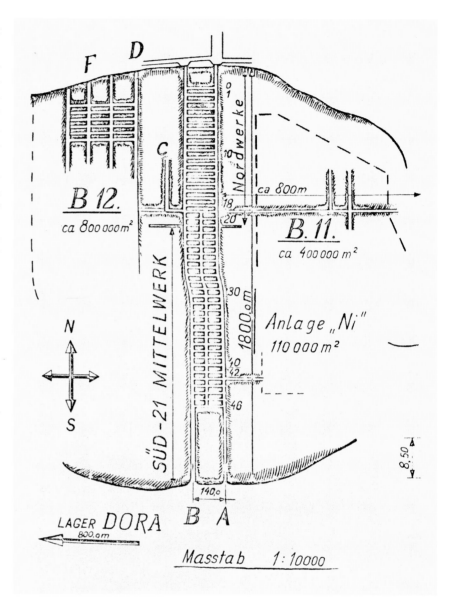

The Mittelwerk factory was engaged in machining components for, and assembly of, flying bombs and rockets. Situated at Niedersachswerfen, near Nordhausen, the factory had 2 equidistant tunnels that passed right through the hill, each taking a standard gauge railway line for the supply of materials and loading of rockets.

and its products, before handing it over to the Red Army. Nordhausen had been allocated to the Soviet zone of Germany, which subsequently became the German Democratic Republic.

Had the V-2 come sooner, its effect on the outcome of the war might indeed have been decisive. As it was, some 1,403 were launched at London, killing 2,754 people and injuring another 6,532. Many more were launched at targets in Belgium in the last months of the war—1,214 at Antwerp alone. Its designers, including Werner von Braun, went to the US after the war and designed rockets for the Western Allies. Combined with a nuclear warhead, the ballistic missile became the ultimate weapon in whose precarious balance of terror the world has existed ever since.

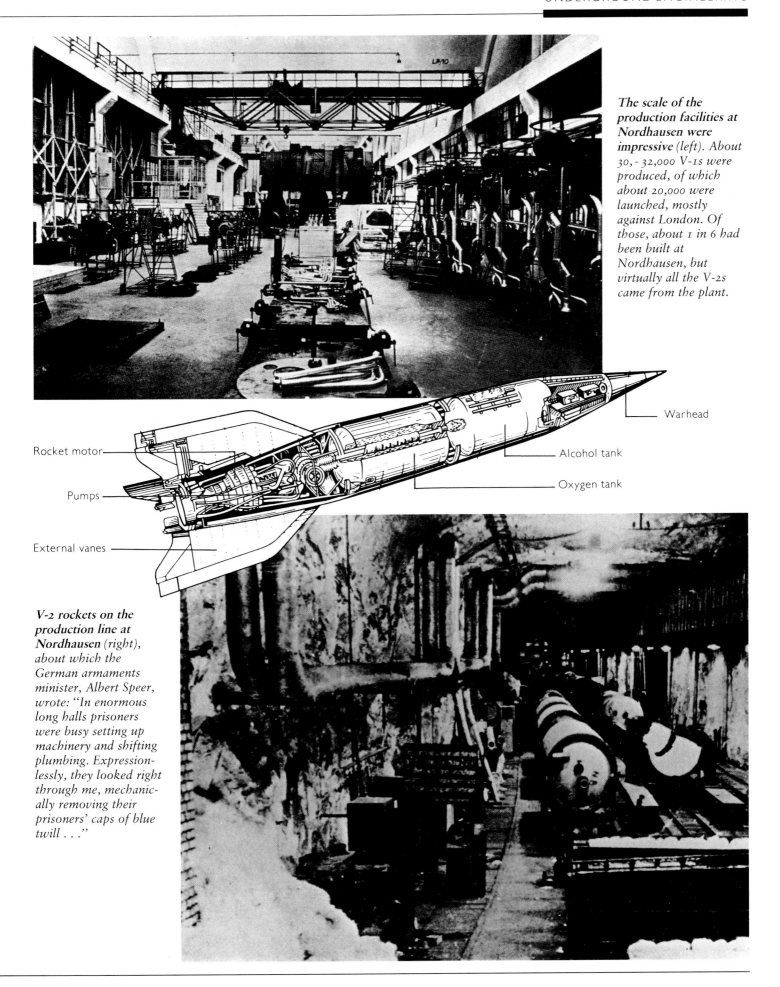

The scale of the production facilities at Nordhausen were impressive (left). About 30,- 32,000 V-1s were produced, of which about 20,000 were launched, mostly against London. Of those, about 1 in 6 had been built at Nordhausen, but virtually all the V-2s came from the plant.

Warhead

Rocket motor

Alcohol tank

Pumps

Oxygen tank

External vanes

V-2 rockets on the production line at Nordhausen (right), about which the German armaments minister, Albert Speer, wrote: "In enormous long halls prisoners were busy setting up machinery and shifting plumbing. Expressionlessly, they looked right through me, mechanically removing their prisoners' caps of blue twill . . ."

Giant Collision Tunnel

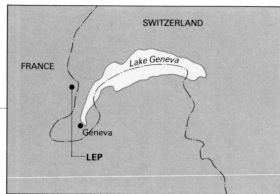

Fact file

The world's largest scientific instrument

Builder: European Council for Nuclear Research (CERN)

Built: 1983–89

Length of tunnel: 17 miles

Cost: £500 million

The scale of machinery (opposite) within LEP can be gauged from the detectors in the experiment halls at the points of collision: the apparatus in L3 is mostly encased in a magnet that contains 6,500 tons of steel; Delphi is equipped with the world's largest super-conducting magnet. The annual budget, shared by 450 physicists from 39 institutions, is close to £200 million.

The world's largest scientific instrument lies in a tunnel 17 miles long on the border between France and Switzerland. The large electron-positron collider, or LEP, is a machine for accelerating sub-microscopic particles to speeds close to the speed of light, before smashing them together and watching the results. The ring is as large in circumference as the Circle Line on the London Underground. Arranged along a tunnel 12 feet in diameter are 4,600 huge magnets to guide the beams of particles through an evacuated tube. More than 60,000 tons of technical equipment have been installed to create a machine that consumes 70 megawatts of electrical energy—enough for a large city. From the surface this huge construction is scarcely visible.

Particle physicists use the biggest machine in the world to study the smallest particles, the fragments of matter from which the entire universe is made. Once it was believed that the atom was the smallest particle that could be created, but scientists have long since shown that atoms themselves consist of yet smaller particles—electrons, protons, neutrons and others. One cannot take atoms apart with a scalpel, or look at them through a microscope; only brute force will break the bonds that hold them together. So the process of discovery has consisted of dissecting atoms by force, and then hurling the constituent parts at one another as hard as possible and watching the bits that fly off.

The greater the relative speed of particles, the more complete the rupture when they collide. Since the first particle accelerators were built in the 1930s, they have become bigger and bigger as the energy—and hence the speed—of the particles has increased. At LEP, two different sorts of particles are accelerated, in different directions, then collided together like two vehicles in a head-on crash.

Electrons, discovered by J.J. Thomson at the University of Manchester during the 1890s, are very light particles carrying a negative charge. Positrons were first found by Carl Anderson at the California Institute of Technology in 1932, and as their name suggests are positively charged. They are in fact identical to electrons except for their charge. The fact that both these particles have electrical charges makes it possible to guide them using magnets, and increase their speed using radio-frequency fields.

What happens at LEP is that the electrons and positrons travel around the evacuated tube of the accelerator in bunches. Because they have opposite charges, the magnetic and electrical fields make them travel in opposite directions until, at the will of the experimenters, they collide. When they do, they annihilate each other, creating for a fraction of a second a burst of high energy that mimics the state of the universe at the moment of its creation. Instantly the energy rematerializes as particles again, but in that fraction of a second the scientists have created the conditions they wish to study.

Accelerators must be big because if particles are made to go round a very tight circle, they lose energy much too fast and slow down. So bigger energies require bigger accelerators, and that means more expensive ones. The point has long been passed when individual universities, or even individual European nations, could afford to build competitive accelerators on their own. So in 1954 twelve European nations got together to form CERN (Conseil européen pour la recherche nucléaire), a collaborative venture in which all shared the costs. Two more nations have since joined and the organization, based in Geneva, has changed its name to the more accurate European Laboratory for Particle Physics. CERN, however, remains the acronym by which it is known around the world.

Planning for LEP began in the late 1970s when it was realized that in order to keep up with the American laboratories CERN would need a new and bigger machine. Approval was given in December 1981, and actual work began in September 1983. The machine produced its first electron-positron collisions in August 1989, less than six years later.

LEP was built underground for a variety of reasons. Doing so provides a stable and secure foundation for a machine that, despite its size, must be positioned and maintained with extraordinary precision. There are, in any case, few areas of land flat enough to create a circle more

Giant Collision Tunnel

ALEPH

Experimental halls

Access points

Point 4

Point 3

Point 2

Point 1

Focussing magnets

Accelerating cavity

Bending magnets

Vacuum chamber

Collision

An aerial view of the LEP site (above), showing the route of the tunnel. The border between France and Switzerland is marked by the dotted line, with Switzerland in the foreground. The small circle within LEP on the left marks the circuit of the Super Proton Synchrotron, one of 3 accelerators built between 1954 and the construction of LEP.

Each collision of electrons and positrons produces enough information to fill a telephone directory, but so sophisticated are the monitoring equipment and high-speed computers that only unusual or interesting data is recorded for analysis. Within the vacuum chamber, the circulating electrons are accelerated, bent and focused.

than 5 miles in diameter without considerable tunnelling or the building of embankments, particularly close to Geneva. And by burying the machine out of sight, CERN made sure of obtaining permission from the local authorities for the project.

The actual site of LEP is a small stretch of land between Lake Geneva and the Jura mountains. The tunnel is almost but not quite circular, consisting of eight straight sections 550 yards long, linked by eight arcs each 3,000 yards long. Most of the tunnel was drilled through a soft rock called molasse, using full-face boring machines. They were lowered to the level of the tunnel through 18 shafts running from the surface to a level between 160 feet and 480 feet below the surface.

The boring machines, guided by a laser beam to ensure they were precisely on course, cut their way through the molasse, and the tunnel behind

was lined with precast concrete rings. The largest excavations in the tunnel are the experimental halls, huge underground caverns where the beams of electrons and positrons collide and instruments are set up to detect the results. A total of 1.83 million cubic yards of spoil was removed and dumped on land of low agricultural value close to the tunnel. Finally topsoil was replaced, and the land returned to farming.

Once the tunnel was complete it was fitted out with the tube around which the particles travel, and the bending and focusing magnets that guide them. Because the particles must not collide with any other matter as they move, the tube containing them must be completely evacuated. The electrons and positrons each travel more than 100 times the distance between the Earth and the Sun as they whirl around, so removing the last traces of air from the vacuum tube is critical. The beam runs through the centre

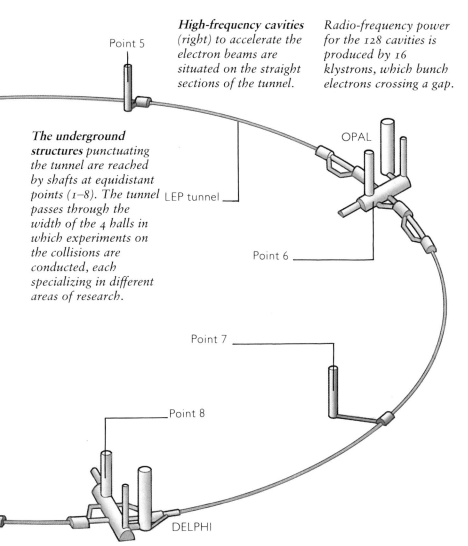

Point 5

High-frequency cavities *(right) to accelerate the electron beams are situated on the straight sections of the tunnel.*

Radio-frequency power for the 128 cavities is produced by 16 klystrons, which bunch electrons crossing a gap.

The underground structures *punctuating the tunnel are reached by shafts at equidistant points (1–8). The tunnel passes through the width of the 4 halls in which experiments on the collisions are conducted, each specializing in different areas of research.*

OPAL

LEP tunnel

Point 6

Point 7

Point 8

DELPHI

of an oval tube with pumps along it to maintain the vacuum.

Arranged along the course of the tube are 3,368 dipole magnets which bend the beam and keep it on course, and 816 quadrupole magnets which focus the beam, keeping it sharp and narrow. In addition, 128 accelerating cavities boost the speed of the particles. In all, this hardware amounted to 14,000 tons of steel and 1,200 tons of aluminium, and it was carried to the appropriate position around the ring using a monorail system. Measuring instruments of extreme precision were used to position the magnets to within four-thousandths of an inch.

The actual detectors used to measure the results of the collisions are themselves spectacular objects, more than 30 feet long and the same in diameter, straddling the pipe where the beam runs. There are four such detectors, designed by huge teams of scientists. The detector for the experiment called L3, for example, is as big as a five-storey building and is literally crammed with electronic instruments.

LEP was finished on time, and cost 1.3 billion Swiss francs (£500 million), just 5 percent more than the initial budget. Within weeks of starting up, it had demonstrated its worth by confirming that all matter is made from just three families of subatomic particles. This proof of what theoretical physicists call the Standard Model was a triumphant vindication of the effort and money put into the world's biggest and most sophisticated machine. It represented another step on the way to the physicists' holy grail, the so-called TOE, or Theory of Everything, which would explain the entire workings of the universe from the smallest particle to the largest heavenly body. When that is achieved, the physicists will have worked themselves out of a job, but none is worrying about that just yet.

The linac injector *(above) supplies the electrons and positrons which are then stored until they are fed into the Proton Synchrotron (PS) followed by the Super PS; these are interconnected accelerators which give the particles their first big boost in energy before they are fed into LEP where they are taken to even higher speeds.*

Railroad under the Sea

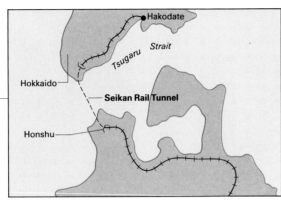

Fact file

The world's longest and most costly tunnel

Builder: Japan Railways

Built: 1971–88

Length: 34 miles

Minimum depth below sea: 275 feet

The world's longest and most expensive tunnel lies clean, dry and hardly used beneath the Tsugaru Strait between the islands of Honshu and Hokkaido in Japan. Only 15 trains a day each way pass through this extraordinary tunnel, 34 miles long and dug through some of the most difficult rock ever encountered. It took twice as long as expected, and cost almost ten times as much. Completing it was "a technological achievement without parallel in the world" according to the Japanese Minister of Transport who opened it in March 1988. But while the tunnellers laboured underground, the Japanese passengers took to the skies. By the time the first train rolled through, air services were already so well established between the two islands that few wanted to travel by train.

The tunnel represents the partial fulfilment of a dream that has inspired Japanese engineers since 1936—to link all the islands of Japan together by train. Originally the line was to go even farther north to the island of Sakhalin, then on to Korea, which was a Japanese colony. But Sakhalin fell to the Soviet Union in World War II, and Korea became independent. So a modi-

The Tsugaru Strait is prone to extreme weather and violent currents that close it for at least 80 days a year. It was the loss of 5 ferries during a typhoon in 1954 that initiated investigations into the feasibility of building a tunnel. In the foreground is the Tappi construction site.

fied plan, linking the four islands of the Japanese archipelago, was adopted, and completed with the Seikan Tunnel and the Seto Bridges, which join the main island of Honshu with Hokkaido to the north and Shikoku to the south. The fourth island, Kyushu, was already linked to Honshu through the Kanmon Tunnel, opened in 1942. Following the loss of five ferries and 1,430 lives during a typhoon in the Tsugaru Strait between Honshu and Hokkaido, which is 15 miles wide at its narrowest point, an investigation was made into the feasibility of a tunnel.

The surveys established that the task would be a difficult one. The rocks of Japan are geologically young, created by volcanic action and full of faults and fissures. They are both unstable and porous, allowing large water flows—the very worst kind of rock to drill through. Treacherous sea conditions in the strait made surveying difficult, so less information was obtained than the engineers of Japanese National Railways had hoped for. In March 1964 the first shaft was drilled on the Hokkaido side, followed two years later by a similar inclined shaft on the Honshu side. The idea of the shafts was to provide a proper survey, develop a method of tunnelling through the rock, and ultimately to serve as entrances to the main tunnel.

The shafts proved that a tunnel could not be drilled without first making the rock less porous and unstable. This was done by the technique known as grouting. Small holes were drilled into the rock ahead of the advancing face, and fanning out in the shape of a cone. Into these holes a mixture of cement and a gelling agent was pumped under huge pressure, so that it was forced into all the small fissures in the rock, sealing them. Then the main tunnel was advanced through the same volume of rock, before drilling and grouting a fresh section ahead. Without such careful preparation, the tunnel workings would have been flooded almost as soon as they began.

Less than half of the 34 miles of tunnel actually lies under the sea, but this section inevitably proved the most difficult. To reduce seepage from the sea, the tunnel was cut more than 300 feet beneath the seabed. For every day's cutting of the tunnel, two to three days might be spent grouting in an attempt to improve the rock. A pilot tunnel was driven ahead of the main and service tunnels, to provide advance warning of difficult conditions, and core samples of rock were taken by a small-bore drill driven into the

rock in advance of the tunnel faces.

In spite of such precautions, there were at least four major floods, the worst occurring in 1976 and 1977, when flow rates of up to 80 tons of water a minute forced the miners to evacuate the tunnel in a hurry. In one of these incidents, in May 1976, more than 2 miles of the service tunnel and 1 mile of the main tunnel were flooded, delaying work for months. Eventually the service tunnel was detoured past the region of difficult rock, and additional grouting and special mining techniques used to get the main tunnel through the same region without further problems. Even today, with the tunnel completed and lined, four separate pumping systems must operate continuously to keep it dry. Without the pumps the tunnel would fill with water in 78 hours.

The nature of the rock made it impossible to use full-face boring machines of the type being used to cut through the chalk beneath the English Channel. Instead, the engineers were forced to use conventional mining methods,

breaking up the rock with explosives before clearing it away with mechanical excavators.

Because of the length of the tunnel, special facilities had to be provided for ventilation and fire prevention. Air is pumped into the tunnel through inclined shafts on either side of the strait, flowing through the pilot tunnel into the centre of the main tunnel and then outward through both outlets of the main tunnel. A steady breeze of about 2 miles an hour is enough to keep the air in the tunnel fresh, and prevent overheating from the heat generated by the trains. In the service tunnels, which run alongside the main tunnel through the sub-sea section, the air pressure is slightly higher, ensuring that air flows from the service tunnel into the main tunnel, and not the other way round. This would be vital in the event of a fire, because it would prevent smoke entering the service tunnels, which are the escape route for passengers.

If a fire did break out, the train's driver would try to take it out of the tunnel as quickly as possible, but because of the length of the tunnel

The single-bore tunnel *has twin 3 foot 6 inch gauge tracks, but it has been built to dimensions that will allow standard gauge (4 feet 8½ inches) Shinkansen trains to use it when the money is available. The delay in building the faster trains has removed the tunnel's raison d'être, since the time saving is not sufficient to make major inroads into alternative modes of transport.*

Railroad under the Sea

that might prove impossible. For this reason two emergency stations have been built underground, where passengers could get off the train and run down escape and rescue passages into the service tunnel. Smoke exhaust blowers are provided at these stations to remove it as quickly as possible, while emergency generators would ensure the tunnel remained well-lit.

Along the tunnel there are four heat-detection systems and sprinklers, designed to provide early warning of a fire and to put it out rapidly. Many of these facilities were designed after a disastrous fire in another Japanese rail tunnel in November 1972 killed 30 people and injured many more. "We think this is the safest undersea tunnel in existence" boasts Mr Shuzo Kitagawa, the Deputy Project Director.

Safe it may well be, profitable it is not. The total cost of the tunnel was $6.5 billion, against an original estimate in 1971 of $783 million. Together with financing costs and other incidentals, the total amounted to $8.3 billion. While the cost went up, the potential use went down. In the ten years from 1975, the number of passengers using the ferries declined by 50 percent. In 1986, before the tunnel opened, only 186,000 travelled by train and ferry between Tokyo and Hokkaido's main town, Sapporo, while 4.5 million—25 times as many—went by air, which takes 90 minutes. The decision not to run the bullet trains through eliminated any chance of reversing this trend, and there was even talk of abandoning the tunnel and using it as an underground reservoir for oil, or even for growing mushrooms. But that would have wounded the pride of Japanese National Railways unacceptably, so the tunnel was completed and brought into use. Debt repayments and operating losses are now put at $67 million a year for the next 30 years.

The daytime trains have made few inroads into the air traffic between Tokyo and Sapporo, but the sleepers have proved more successful, and three 12-car trains a night pass in each direction. Travellers on the route can watch illuminated displays in each coach which chart the progress of the train through the tunnel, showing distance covered, and depth below the sea; the maximum depth reached is 787 feet. Many passengers spend a good part of the journey taking pictures of these panels, and themselves. A stop is also made in the middle of the tunnel for two minutes to enable passengers to take pictures through the windows of panels on the tunnel wall.

Mechanical excavation (right) was employed in places but generally explosives were used, after which the surface was stabilized with a cement mixture called shotcrete and then lined with H-section steel supports and a layer of concrete 2 feet thick, or 3 feet in weak ground.

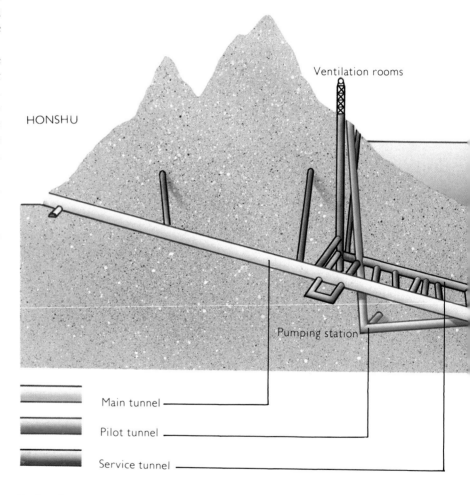

Ventilation rooms

HONSHU

Pumping station

Main tunnel

Pilot tunnel

Service tunnel

Shafts of two types connected the Tappi and Yoshioka construction sites with the tunnels: the vertical shafts were used to transport machinery, materials and shotcrete, and personnel working in the main and service tunnels; the inclined shafts conveyed workers to the pilot tunnel and was the conduit for the largest machinery, and extracted water and spoil. The inclined shafts formed the air intakes and were adapted for the forced air ventilation system, as well as providing a passage for maintenance, escape and rescue. The vertical shaft took exhaust air from the tunnel and is now a smoke exhaust shaft should fire break out in the tunnel.

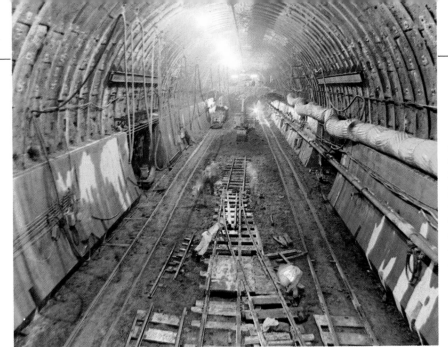

A narrow gauge railway was used in the main tunnel to bring grouting, water, concrete, grout pumps and power units to the headings and to extract spoil. A fire alarm system, radio telephone communication and loudspeakers in the tunnels were part of the safety measures.

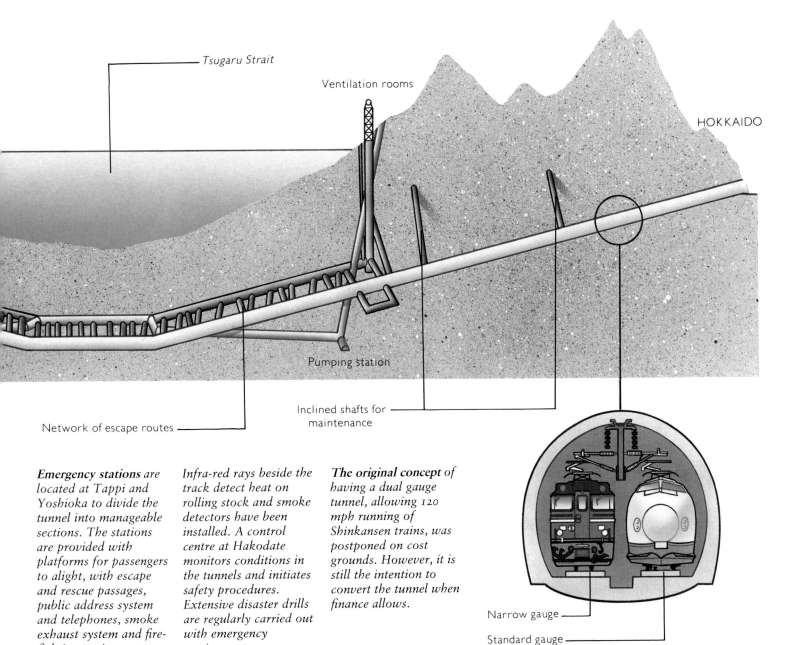

Tsugaru Strait

Ventilation rooms

HOKKAIDO

Pumping station

Network of escape routes

Inclined shafts for maintenance

Narrow gauge

Standard gauge

Emergency stations are located at Tappi and Yoshioka to divide the tunnel into manageable sections. The stations are provided with platforms for passengers to alight, with escape and rescue passages, public address system and telephones, smoke exhaust system and fire-fighting equipment.

Infra-red rays beside the track detect heat on rolling stock and smoke detectors have been installed. A control centre at Hakodate monitors conditions in the tunnels and initiates safety procedures. Extensive disaster drills are regularly carried out with emergency services.

The original concept of having a dual gauge tunnel, allowing 120 mph running of Shinkansen trains, was postponed on cost grounds. However, it is still the intention to convert the tunnel when finance allows.

Great Tunnels

The earliest tunnels were built in the tombs of Babylonian and Egyptian kings. A small tunnel is believed to have been built under the River Euphrates in the twenty-second century BC.

The mining of minerals and the walls of castles under siege kept alive the skills of tunnelling until the canal age of the eighteenth century, when the work of engineers such as James Brindley eclipsed all that had gone before. With the railways came the invention of the tunnelling shield and the use of compressed air, both to counteract the external pressure of water and to power drills.

The major mountain massifs on important routes have been bored. The future of tunnelling lies with projects such as the Channel Tunnel, which will just exceed in length the comparable Seikan Tunnel.

The Malpas Tunnel, Canal du Midi

Although comparatively short, at 528 feet, the Malpas Tunnel in south-west France is a tunnel with more "firsts" than any other: it was the first canal tunnel to be built, completed in 1681; it was also the first tunnel to be built for any form of transport; and it was the first tunnel to be excavated with the help of gunpowder, which represented a major breakthrough in technique. The Canal du Midi was the first great European canal, linking the Atlantic Ocean with the Mediterranean Sea. In its 148 miles there are 119 locks to take the canal over a summit of 620 feet above sea level.

The Rove Tunnel, Canal de Marseille au Rhône

The world's longest canal tunnel, at almost 4½ miles, was opened in 1927 to link the port of Marseilles with the Rhône at Arles. Built to take seagoing vessels, the tunnel is 72 feet wide and 37 feet high. Work began on the tunnel at the south end in 1911 and at the north end in 1914. The outbreak of World War I halted work until German prisoners of war were assigned to the tunnel. In 1916 they broke through but work on lining it was slow. Twice as much spoil was excavated from the Rove as the Simplon Tunnel, due to the size of the bore. A collapse of a section of the tunnel in 1963 has closed it to traffic.

The Channel Tunnel, Folkestone–Sangatte

The idea for a tunnel under the English Channel is almost 2 centuries old: the first proposal, in 1802, by a French mining engineer was followed by a number of schemes over the next 160 years; all were rejected by Britain for fear of jeopardizing national security. A scheme in the 1870s and '80s produced pilot tunnels from both sides before the British government put a stop to the digging. Work began again in 1973, only to cease on the grounds of economic stringency.

In January 1986 Eurotunnel was awarded the concession to build the present rail tunnel, which will comprise 2 single bores with a service tunnel between. Of the total length of 30.7 miles, 23.6 will be under the sea. The saturated chalk through which the tunnellers have to bore poses the main challenge. A seal between the cutting head of the Japanese boring machines and the cylinder behind it prevents water flooding in. The tunnel lining sections are bolted together inside the cylinder, and the cavity left behind as the cylinder moves forward is filled with compressed concrete. The tunnel will transform the prospects for rail transport in Britain and become part of the European high-speed rail network.

CHAPTER FIVE

Astronomical Constructions

Astronomers are at the edge of the greatest leap forward in understanding the Universe since the first telescopes were produced in the seventeenth century. Those primitive instruments revealed far more objects in the sky than could be seen by the naked eye, but for every new object they identified, a thousand more remained hidden. The limitations of the telescopes, and the fuzzing of the image brought about by the Earth's atmosphere, have proved frustrating obstacles to understanding.

The most obvious way around the difficulties was to build better telescopes, and site them, like those of the European Southern Observatory, high on a mountain peak where the atmosphere is thinnest. Isaac Newton showed the way, with telescopes based on mirrors rather than lenses. Since then, the mirrors have become larger and larger, capturing ever greater quantities of light so as to detect even the faintest of stars.

But to begin observations beyond the atmosphere altogether, as the Hubble Space Telescope will do, is to open new doors. It will make possible the study of individual stars that have previously been an indistinct part of a cluster. Objects infinitely remote and impossibly dim will for the first time be seen by the human eye. The findings will be combined with the results now available from radio telescopes of huge size and great sensitivity, which detect the signals emitted by exotic objects such as quasars and pulsars. Foremost among these is the Very Large Array, a dull name for a remarkable instrument constructed on a huge plateau in New Mexico.

The search for knowledge has driven scientists in many disciplines to build ever bigger and more expensive machines. Nowhere is this more likely to produce dramatic discoveries than in astronomy, which could be set for its most productive decade since the 1920s.

Astronomical Constructions
Space Telescope
Very Large Array
European Southern Observatory

Cosmic Time Machine

Fact file

The world's most powerful gatherer of information about the Universe

Coordination agency:
Marshall Space Flight Center, Huntsville, USA

Built: 1977–85

Length: 43 feet

Diameter: 14 feet

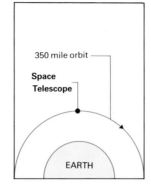

350 mile orbit

Space Telescope

EARTH

A complex satellite the size of a railroad car spent five years waiting for the opportunity to transform our picture of the Universe, its launch delayed by the Challenger disaster in 1986. The Hubble Space Telescope spent the interim period under intensive tests, and safely stored in a clean room at the Kennedy Space Center in Florida, wrapped in a plastic bag with clean air blown through it. Its launch on 24 April 1990 represented the beginning of the greatest opportunity for discovering new information about the Universe since Galileo directed the first simple telescope at the skies almost 400 years ago.

The Space Telescope will provide the clearest and deepest view of the Universe astronomers have ever enjoyed. Floating above the disturbing effects of the atmosphere, it will be able to observe the heavens using ultraviolet and infrared rays as well as visible light. It will pick up objects far too distant or too faint to be seen on Earth, so remote that light from them takes billions of years to reach us. By seeing so far, the Space Telescope will look backward into the past, to events that took place 14 billion years ago when the Universe was young. It will see objects 25 times fainter than anything visible from Earth, and explore the Universe in detail ten times greater than ever previously achieved.

Astronomers have known for decades that the view of the Universe from above the atmosphere would be much clearer than it is from Earth. The twinkling of the countless stars in the sky is caused by atmospheric disturbance bending the waves of light that reach us. Looking at stars from the ground is like viewing birds flying overhead from the bottom of a swimming pool. So the first proposals for a space-based telescope are older even than space travel itself. They came in 1923 from Hermann Oberth, the German pioneer who developed many of the key concepts in space exploration.

In 1962 the US National Academy of Sciences recommended developing a large space telescope, an option supported by similar groups in 1965 and 1969. The launch of space observation satellites by the US National Aeronautics and Space Administration in 1968 and 1972 further whetted the appetite, but it took the development of the Space Shuttle to provide the means for launching a really large telescope into space. The European Space Agency became involved in 1975, funding was finally in place by 1977, and the telescope was ready by 1985.

The Space Telescope is essentially similar to a large optical telescope here on Earth. Unlike Galileo's pioneering instrument, modern telescopes use mirrors rather than lenses to focus the light. The very biggest have mirrors 200 inches in diameter, in order to collect light from the widest possible aperture, and hence detect the most distant of objects. The Space Telescope's mirror is 94 inches in diameter, and made of a special type of glass which expands and contracts very little with changes of temperature. It took the Perkin-Elmer Corporation 4 million person-hours of work to cast and polish it. The finish they achieved is staggering. The curvature is accurate to within two-millionths of an inch;

The Space Shuttle is central to the concept of the Space Telescope: Discovery has placed the telescope in orbit and a Shuttle will carry astronauts to service it every 5 years.

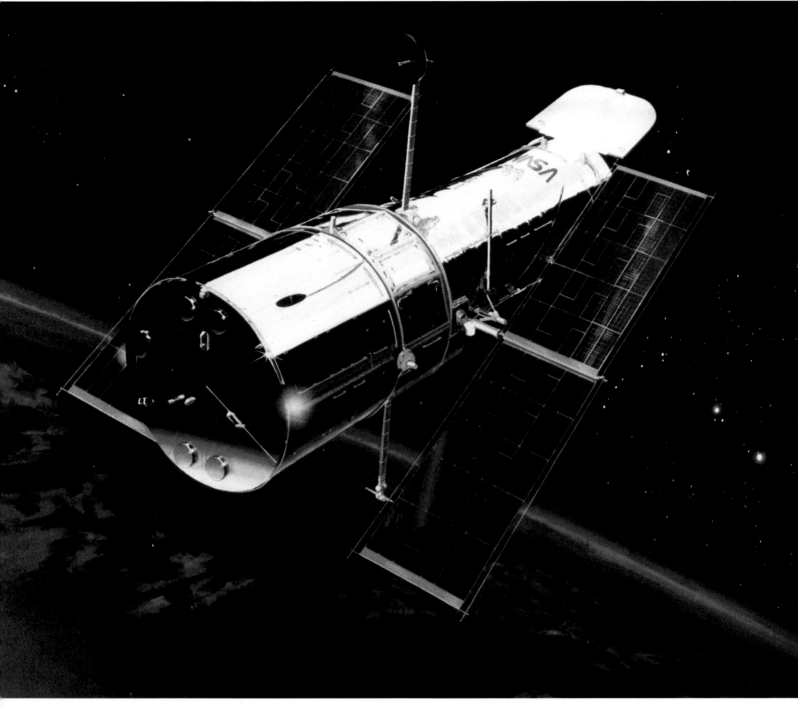

that means that if it were scaled up to the size of the Earth, no irregularity on its surface would be greater than 12 inches. Its resolution—the ability to separate two distant objects—could distinguish the headlamps of a car 2,500 miles away, a higher resolution than any other astronomical telescope.

The mirror glass is covered with an aluminium reflecting surface. Mounted well inside the cylindrical body of the telescope, it will reflect light forward to a second mirror 12 inches across and 16 feet farther forward. This mirror will return the light through a 2-foot hole in the middle of the primary mirror to the focal plane where the scientific instruments are mounted.

There are five such instruments: two cameras, two spectrographs, and a photometer. The Wide Field Camera will be used for investigating the age of the Universe and searching for new planetary systems around young stars. It will be able to see Halley's Comet, normally visible only when, once every 75 years, it comes within close proximity of the Earth. Despite the camera's name, the width of its view will actually be very limited, only 2.67 arc seconds, so that it would take a montage of 100 images to get a picture of the entire Moon. But this narrow view gives far better resolution of distant objects.

The second camera, the Faint Object Camera, will have an even narrower gaze, only a fortieth

The telescope's mirror pointing system is so accurate that it could fire a laser over 400 miles and hit an object the size of a coin, despite the fact that it is moving around the Earth at a speed of 17,000 mph.

Cosmic Time Machine

The telescope's covering is several layers of shining metallic foil which reflects most of the sunlight and prevents overheating. Electrical power is provided by 2 solar arrays, each wing containing 24,000 solar cells, supplemented by 6 batteries to store electricity while the satellite is hidden from the Sun by the Earth.

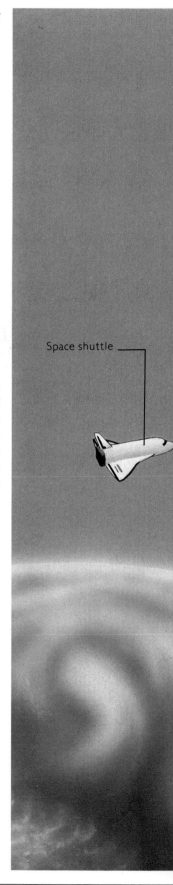

Space shuttle

as great as the Wide Field Camera, but will be able to extend the telescope's range to the limits and produce the sharpest views. Many objects barely visible from Earth will appear as blazing sources of light to this camera.

The two spectrographs will be used for analysing the spectra of light from the objects viewed. From the various characteristic wavelengths given off by different atoms, spectrographs enable astronomers to determine precisely what elements are present in the bright objects they are looking at. The Faint Object Spectrograph will be used for studying the chemical properties of comets, or comparing the composition of galaxies close to the Earth with those far away.

The High Resolution Spectrograph will be used for studies of the chemical composition, temperature and density of the gas that fills the space between the stars, and for studying the atmospheres of planets in our own solar system.

The final instrument, the High Speed Photometer, will measure the brightness of the objects in space, looking for clues that black holes actually exist and providing an accurate map of the magnitude of stars.

All this will be done with an instrument 43 feet long and 14 feet in diameter, weighing just over 11 tons. The telescope is wrapped in several layers of shiny metallic foil, which reflects most of the sunlight and prevents overheating. The Shuttle has placed the telescope in a 380-mile orbit, and the two remained alongside for a couple of days to enable the Shuttle to check that all the systems were working well before return-

ing to Earth. Several months' work by ground controllers were subsequently needed to activate all the systems on the telescope, align the mirrors and verify its exact orbit.

Should anything go wrong, the telescope is designed to be serviced in orbit by the Shuttle throughout its 15-year life. It is expected that several components will need replacing during that time, and it is possible that better scientific instruments may be made to replace the existing ones. The design of the telescope means that individual systems can be pulled out and a replacement plugged in without disturbing anything else. The Shuttle will rendezvous with the telescope, pull it into its cargo bay, and the astronauts will put on space suits and go outside to service it. Regular servicing should keep the telescope working well.

The name of the Space Telescope comes from that of a famous American astronomer, Edwin Hubble, born in 1889. Hubble used the big optical telescopes of his day to make many important discoveries. He proved that many of the objects we can see in the sky are not in our galaxy at all, but are galaxies of their own, millions and billions of light years away. He also showed that the entire Universe is expanding.

One of the most important results the Space Telescope is expected to obtain is the value of the Hubble constant, which describes the rate of expansion and the age of the Universe. The work of Edwin Hubble in the 1920s transformed our understanding of the Universe we live in. The Space Telescope named after him is quite likely to do the same.

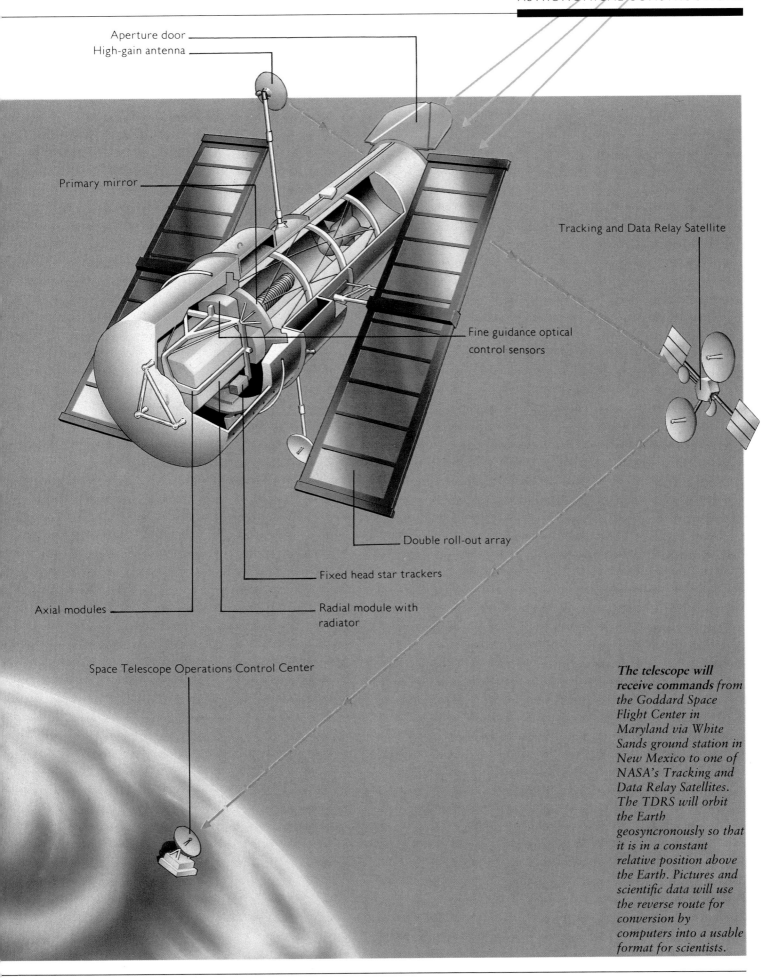

Aperture door

High-gain antenna

Primary mirror

Tracking and Data Relay Satellite

Fine guidance optical control sensors

Double roll-out array

Fixed head star trackers

Axial modules

Radial module with radiator

Space Telescope Operations Control Center

The telescope will receive commands from the Goddard Space Flight Center in Maryland via White Sands ground station in New Mexico to one of NASA's Tracking and Data Relay Satellites. The TDRS will orbit the Earth geosyncronously so that it is in a constant relative position above the Earth. Pictures and scientific data will use the reverse route for conversion by computers into a usable format for scientists.

The Ultimate Radio Telescope

The world's most powerful radio telescope is to be found on a high, flat plain in New Mexico. From this remote and quiet spot it scans the heavens, producing radio images of stars, galaxies and other exotic objects as sharp as the photographs from the best optical telescopes. Its detailed pictures of some of the millions of objects in the sky that emit radio waves are used to try to understand the huge forces at work: curved filaments a million light years long; objects so dense that light itself cannot escape from them; thin channels across space carrying huge quantities of energy; and the faint blush of radiation that mantles the sky—the last vanishing echo of the Big Bang which began it all 15 billion years ago.

Radio astronomy was invented in the 1930s by an engineer at Bell Telephone Laboratories, Karl Jansky, who was trying to find the source of the crackles and hisses that interfered with transatlantic radio transmissions. Using primitive apparatus, he established that the radio signals came from the heart of the Milky Way. After World War II, more sensitive instruments were built to attempt to point more accurately at radio sources in the sky, to see if they were identical to visible objects, and to resolve their fine structure. Because the signals are so faint compared with terrestrial radio waves, big dish-shaped antennae are needed to detect them, and the first, the 250-foot Jodrell Bank telescope in Cheshire, England, was completed in 1957.

The bigger the dish, the greater the sensitivity and the more accurately it can pinpoint objects in space. But there is a limit to how big a dish can be built, particularly when it has to be steered and pointed with exquisite accuracy. From quite early in the history of radio astronomy it was realized that the effect of a very large dish could be simulated by combining the signals from several smaller ones some distance apart.

The Very Large Array (VLA) in New Mexico is the ultimate expression to date of this type of telescope. One of four telescopes run by the US National Radio Astronomy Observatory, it consists of 27 dishes, arranged alongside three straight rail tracks fanning outward from a central point. Each dish is 82 feet in diameter, its parabolic surface formed from aluminium panels accurate to 20 thousands of an inch over their whole surface. The movable part of the antenna, the dish itself, weighs 100 tons, while the whole structure weighs 235 tons. Each of the 27 dishes can be pointed anywhere in the sky

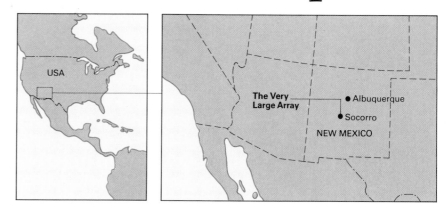

with an accuracy of 20 seconds of arc—equivalent to one 180th of the diameter of the Moon.

Two of the three tracks running outward are 13 miles long, while the third measures just under 12 miles. Each has nine dishes arranged along it. The tracks make it possible to move the dishes in and out, which achieves the same effect as the zoom lens on a camera. Four different configurations of the dishes are possible. In the A configuration, they are distributed along the full lengths of the tracks, giving the greatest resolution for observing small, intense radio sources. In the D configuration, they all cluster within 650 yards of the centre, which enables large, diffuse sources to be studied with lower resolution but higher sensitivity. Configurations B and C are intermediate settings.

To move a dish, a special transporter picks it from its pedestal, carries it to the nearby track, transports it to a different position, then sets it down on another pedestal with a precision greater than a quarter of an inch. Using a pair of transporters, it takes about two weeks to reconfigure the telescope completely, and the process is carried out on a cycle in which the VLA returns to its original configuration every 15 months.

The first design studies for the VLA were drawn up in the early 1960s, and a formal proposal from the NRAO for funds made in 1967. Approval from Congress in 1972 enabled work to begin in 1974, and the whole array was finally completed in 1981. Each dish cost $1.15 million, and the whole telescope was built for $78.6 million, very close to the estimates.

The site chosen, 7,000 feet up on the plains of San Augustin 50 miles west of Socorro, New Mexico, was ideal. Because of the height and the desert climate, there are few clouds to blur the images. Mountains around the site cut out interference from radio, TV and military bases. The terrain is flat, making it easy to move the

Fact file

The world's most powerful radio telescope

Builder: US National Radio Astronomy Observatory

Material: Aluminium

Diameter of dishes: 82 feet

Weight of dishes: 100 tons

All 27 dishes point to the same object in the sky, in a similar fashion to other multi-dish radio telescopes. The VLA represents a quantum leap from the first 2-dish telescope built by the California Institute of Technology near Bishop, California, in 1959. A 3-dish array built at Cambridge, England, in 1963 discovered in 1967 pulsars—radio sources that produce bursts of emissions every few seconds.

The Ultimate Radio Telescope

dishes around, and the site is far enough south to be able to see 85 percent of the sky.

Radio waves from all 27 dishes are amplified a million times by receivers and fed into an underground waveguide which runs along the tracks to the control room at the centre of the array. It is here that the signals from the 27 dishes are combined to create the image. Because some of the waves have travelled farther, from the outlying dishes, their signals will be delayed by a tiny fraction of a second, which is enough to destroy the image. To make the necessary correction, the signals from some of the dishes are automatically delayed before they are all combined in the correlator and passed to a computer for analysis.

The calibration computer first analyses the data for any defects, which might include stray signals from satellites or radar installations. These are removed automatically, and the cleaned-up signals combined using a mathematical technique known as a Fourier transform. The effect is to convert the signals into an image in much the same way that a lens converts light into a picture. The radio image is stored on the computer as a sequence of numbers on a grid, each number representing the strength of the signal, and each grid point a position in the sky. Transmitted to a TV screen, the numbers are displayed as a picture.

Further processing is necessary to produce a crisp image. First, there is a lot of "noise" from stray radio signals that appears like a shower of snow on the picture. This can be removed by averaging the signals over several hours of observation. Then there are distortions arising from the fact that the 27 dishes cover only a small

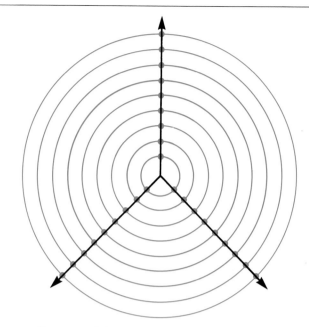

Radio waves (left) are reflected off the curved surface of the dish on to a second subreflector at the focus of the dish, and then to the radio receivers at its centre.

The 27 dishes arranged on the 3 railway lines are connected electronically to synthesize their signals, which are equivalent to those of a single radio telescope 20 miles in diameter.

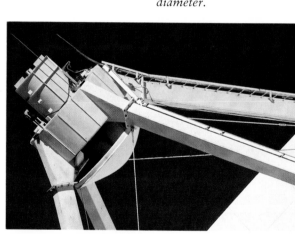

Slight rotation of the subreflector (above), which is an asymmetric mirror, enables signals to be directed to one of 6 receivers tuned to different wavelengths. In the base of the dish are the azimuth and altitude driving motors and gears, which move throughout an observation to compensate for the Earth's rotation.

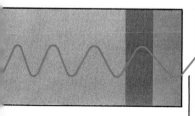

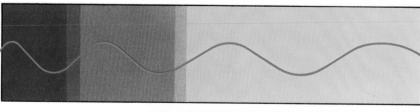

Visible Light

Electromagnetic waves (above) range from gamma rays at the shortest wavelength end of the spectrum, at the left, to low-frequency long waves at the right, with visible light and the colour spectrum near the middle.

Cassiopeia A (right), whose ring of radio emission represents a spherical, expanding cloud of gas expelled by an explosion.

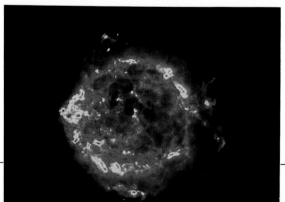

The synchronizing of signals from the 27 dishes is one of the VLA's most vital processes, for the travel time of the signals—from their source to the correlator—must not differ from each other by more than about one 1,000,000,000th of a second. The distance of many miles between dishes can cause a difference in travel time of one 10,000th of a second.

fraction of the area of the array. These can be corrected, to produce an image exactly as it would be seen by a dish the whole area of the VLA. Finally, there is the radio equivalent of the "twinkling" of stars, caused by atmospheric conditions, which can also be corrected for. The final data is stored on magnetic tape, and the images can be further enhanced by taking the tapes to supercomputers elsewhere and running them through special programs.

The VLA operates every day, 24 hours a day, except for a few holidays such as Thanksgiving and Christmas. Astronomers from the US and abroad put in requests to use the instrument, which are considered by a committee and assigned a time. The average proposal requires about eight hours of observing time, although some are completed in as little as five minutes and others need up to 100 hours. Astronomers arrive a day before their observations are scheduled, make sure all is well, and prepare a list of the sources to be observed, for how long, and at what wavelengths. The actual observations are controlled by a computer. Once they are com-

plete, the astronomers depart with their magnetic tapes back to their own laboratories to analyse, and later publish, the results.

More than 700 astronomers use the VLA every year and the images it produces are far from the fuzzy patterns of earlier radio telescopes. They show objects unimaginably far away and driven by forces so huge it is difficult to comprehend them. The radio waves from the source Cygnus A, a galaxy consisting of billions of stars, have taken 600 million years to reach us. Analysed by the VLA, they show two clouds on either side of the galaxy, caused by electrons trapped by magnetic fields. The electrons themselves are speeding away from a bright spot at the very centre of the galaxy.

More dramatic still is the radio source Cassiopieia A, a huge boiling mass of material expanding at a speed of 10 million miles an hour away from the centre of a gigantic supernova that exploded in 1680. Or, to be accurate, that is when we on Earth saw it exploding. The actual explosion took place 9,000 years earlier, for that is how long it took the signals to reach us.

Recorder of the Heavens

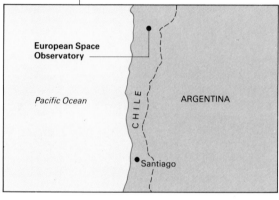

Fact file

The source of the world's best pictures of the sky

Builder: European Southern Observatory

Built: 1964–

Altitude: 7,800 feet

No of telescopes: 15

On a high mountaintop in the Atacama Desert of Chile, 375 miles north of Santiago, is a series of buildings as strange as anything left by the Incas or the Aztecs in Latin America. Dotted around the slopes are 15 telescopes, housed in shiny silver-coloured buildings with white domes. Here astronomers from eight European countries repair to watch the night skies far from the atmospheric pollution and all-pervading light which have made optical astronomy in Europe increasingly difficult.

Observing the skies from the surface of the Earth has certain intrinsic drawbacks. The atmosphere, even when clean, introduces blurring into the images as the light from stars passes through it. The old Royal Observatory established at Greenwich in 1675 (now a London suburb) was, by 1850, suffering from the drawbacks of its location.

Today nobody would dream of building a telescope in a city, or even close to one. Britain's best telescopes are now based on a mountain in the Canary Islands; the major observatory in the US is at Kitt Peak, in Arizona, almost 7,000 feet above sea level. And the nations belonging to ESO—Belgium, Denmark, West Germany, France, Italy, the Netherlands, Sweden and Switzerland—have chosen La Silla (The Saddle), a long ridge 7,800 feet above sea level in the empty vastness of the Atacama Desert.

Because the site is high, the air is thin and there is less of it above the telescopes for the light from stars and galaxies to pass through, so there is much less blurring. Rain and cloud are exceptionally unusual, and La Silla enjoys more than 300 clear nights every year. There is relatively little difference between day and night temperatures, a great advantage because problems can be caused by the expansion and contraction of instruments as the temperature rises and falls.

A site in the southern hemisphere is also highly desirable because so much optical astronomy has, until recent times, been conducted in the north. The south gives the best views towards the centre of our own galaxy and of the Magellanic Clouds, two companion galaxies to our own. The combination of all these features makes La Silla one of the world's finest observatories, where in the opinion of astronomers the best and sharpest pictures of the sky are taken.

Of the 15 telescopes at La Silla, all but one are optical instruments. Eight are financed by ESO, the rest by member countries. The odd one out is a radio telescope, SEST (the Swedish-ESO Submillimetre Telescope), a 48-foot dish which has been listening to very short wavelength radio signals from space since 1987. This instrument is especially well sited because such wavelengths are usually absorbed by water vapour in the European atmosphere. At La Silla it can thrive in the dry air, making observations of molecules in the space between stars in our galaxy and neighbouring ones.

The most interesting telescope at La Silla is undoubtedly the 136-inch New Technology Telescope, which came into use in March 1989. This instrument uses new techniques to produce the best images of many objects in the sky. It is the most advanced land-based telescope in use.

All large modern telescopes are based on concave mirrors made of glass. Casting and polishing such huge pieces of glass to the necessary accuracy is an art form in itself. The glass from which they are made is heated to 2,600°F (1,600°C), and then allowed to cool immensely slowly—up to six months—to avoid stresses. It can then take the best part of a year for further heat treatments, and several years to grind and polish. The mirrors produced are thick and immensely heavy, which causes distortions in the structure and even the shape of the mirror itself as its position alters. A mirror that is perfectly shaped when lying on its back will not

Recorder of the Heavens

The clarity of air around La Silla is helped by ESO's ownership of 300 square miles around the site. It was bought in 1964 to prevent any development, such as mining, that might damage the crystal clarity of the air. There are no cities to cast a glow at night or create atmospheric pollution, the two major problems for European and American observatories.

be quite so perfect when it is held at an angle.

The mirror is made from low-expansion "Zerodur" glass ceramic, and is only half the thickness of a conventional mirror of the same size—less than 10 inches thick. This halves its weight from 12 tons to only 6, reducing the distortion of the supporting structure. In order to eliminate distortion of the mirror itself, which would otherwise be unacceptably large in such a thin piece of glass, it is supported by a unique "active optics" system.

Under the mirror are 78 supports, controlled individually by computer. Light from a star is analysed, and its deviation from the ideal shape—the diffuseness of the image—generates signals which activate the supports, altering the shape of the mirror until a perfect image is obtained. In this way, the mirror always has the right shape, whatever its position. The distortions may be only a few millionths of an inch, but correcting them makes a lot of difference, producing images of stars or other objects at least three times sharper than any other ground-based telescope of the same size.

The NTT is mounted on a thin layer of oil, only a thousandth of an inch thick, which enables it to rotate about a vertical axis. The temperature of the oil is controlled to within 0.1°C to avoid the generation of heat which

The biggest of the conventional refracting telescopes at La Silla is a 140-inch instrument commissioned in 1975. It occupies the biggest dome on the highest point of the site (right). The NTT has eclipsed it for finely resolved pictures.

would disturb the atmosphere around the instrument. The building in which it is housed is, like all observatories, unheated to minimize disturbances. The pointing accuracy of the NTT is ensured by computers that take into account the minute flexing of the structure and allow for it. As a result, the NTT can be pointed in the right direction more precisely than any other telescope of the same size.

The result of all this has been to produce the best images of stars and galaxies ever taken. But the NTT by no means represents the last word in telescope technology. Ahead lie bigger and even more sophisticated instruments, with mirrors 394 inches or more across. The first of these, an

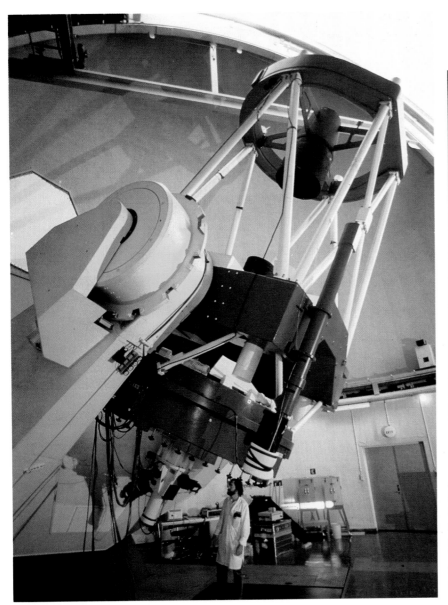

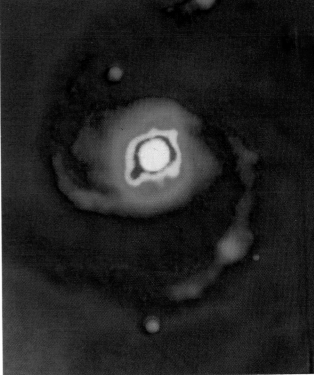

A spiral galaxy *illustrates the clarity of image achievable with the sophisticated technology of the ESO. The NTT is capable of resolving, for the first time, globular clusters at a distance of 700,000 light-years, making possible the study of individual stars. In turn this allows more precise dating of clusters and the study of gas- and dust-clouds.*

The immense size and weight *of the larger telescopes at La Silla demand high-precision mechanics to compensate for the Earth's rotation, particularly when taking a long exposure. The NTT is equipped with computerized position encoders which achieve the greatest pointing accuracy of any ground-based telescope of comparable size.*

American instrument named after its sponsor, William Keck, is due to come into operation on Mauna Kea, 13,600 feet up in Hawaii, in 1991. The Americans are also planning their own National New Technology Telescope, a 645-inch instrument planned for 1992. This should have the capacity to identify objects the size of a gold sovereign 1,000 miles away.

ESO's next step is to develop a system of "adaptive optics" designed to correct for the distortions of the atmosphere. This will be done by placing a small mirror in the telescope, and monitoring the image on it of a bright star in the field of view. Then the mirror will be automatically and rapidly deformed so as to correct any changes in the star's image caused by atmospheric disturbance. The result will be to correct also the rest of the image, producing unprecedented clarity. Tests have shown that the method works, and the next stage is to implement it in telescopes such as the NTT.

Further in the future is the next ESO project—the Very Large Telescope, or VLT, for which $232 million has already been budgeted. Once in operation by 1998, the VLT should be more powerful than all of the 20 largest telescopes currently in use. It will in fact be four telescopes, each with 320-inch mirrors, whose signals can be combined to create the same effect as a single mirror 640 inches across.

The VLT will use the same technology as the NTT, and should be sensitive enough to pick up light from objects 18 billion light years away. Work began on the first of the telescopes in 1988, although a final site for it has yet to be decided. It will be somewhere in the Atacama Desert, but not at La Silla itself. When finished, it will be possible to operate it entirely by remote control from the headquarters of ESO near Munich, without actually having to travel to Chile at all.

Gazetteer

MONOLITHIC MEMORIALS

The Americas

Crazy Horse Memorial, Custer, South Dakota, USA

Situated 17 miles from Mount Rushmore, a tribute to the Indians of North America is being carved out of the granite of Thunderhead Mountain. This colossal project, which will create the world's largest statue, was begun in 1947 by the sculptor Korczak Ziolkowski, who assisted Gutzon Borglum on Mount Rushmore. The site was chosen in 1940 by Ziolkowski and the son of the sculpture's subject, Crazy Horse, who defeated General Custer at Little Big Horn in 1876. Crazy Horse was killed by an American soldier in the following year while under a flag of truce. The sculpture depicts the chief astride a pony, and when complete will be 563 feet high and 641 feet long.

Gateway Arch, St Louis, Missouri, USA

This giant catenary arch, standing on the bank of the Mississippi River, was built in 1966 to symbolize St Louis' position as the gateway to the West. Designed by Eero Sarinen, it is a double-walled stressed-skin structure 630 feet high. The outer wall is made of $\frac{1}{4}$-inch stainless steel, the inner wall of mild steel almost $\frac{2}{5}$ inch thick; the lower part of the gap between them is filled with concrete, the upper with cellular stiffening. In cross-section, the arch is a hollow equilateral triangle, within which lifts travel to an observation platform at the top.

San Jacinto Column, near Houston, Texas, USA

This column, at 570 feet the tallest in the world, was built in 1936–39 on the bank of the San Jacinto River to commemorate the battle that took place there in 1836 between the Texans, under Sam Houston, and Mexican troops. The concrete column, faced with creamy marble limestone, is 47 feet square at the base and tapers to 30 feet square at the observation platform. On top is a vast star, weighing nearly 197 tons.

Europe

Hadrian's Wall, Cumbria and Northumberland, England

The Romans' principal defence in Britain against invasion from the north by warlike Picts and Scots was Hadrian's Wall. Built in 122–30 at the Emperor's command, it ran between natural strongpoints from the Solway Firth, in the west, where it was made of turf, to the estuary of the Tyne in the east, where it was a grey stone structure up to 14 feet high. Along its 73 miles were forts, milecastles and signal turrets, manned by some 18,000 troops.

The wall was protected to the north by a ditch 27 feet wide and 9 feet deep, while to the south was the vallum, a 20-foot-wide, flat-bottomed ditch running between turf walls 10 feet high that served as a road. The Romans abandoned the wall in 383, when Rome itself was attacked by the Goths, but a significant part of it and 17 forts, particularly the well-preserved fort of Vercovium near Housesteads, can still be seen.

Stonehenge, Salisbury Plain, Wiltshire, England

Building of this megalithic monument began c3500 BC, before Egypt's pyramids, and continued for about 1,500 years. It was probably always the site of some form of religious ritual, but it might also have been a primitive astronomical observatory.

The final "building", whose ruins remain, consisted of a ring of sarsen stone monoliths 16 feet high and weighing up to 26 tons, linked by a continuous lintel. This ring encloses a circle of 4-ton bluestones, brought from the Welsh Preseli Mountains 200 miles away, a horseshoe of five sarsen trilithons and a horseshoe of bluestones. In the middle is the great Welsh blue-green sandstone "Altar Stone". The trilithons are made up of two standing stones and a linking lintel, fitted together with precisely shaped ball-and-socket joints.

ARCHITECTURAL ACHIEVEMENTS

The Americas

Boeing Factory, Everett, Seattle, Washington, USA

Located on the outskirts of Seattle, the Boeing Company's manufacturing facilities at Everett are the largest in the world. When it was completed in 1968, the capacity was 200 million cubic feet (70 million cubic feet more than the Vertical Assembly Building at the Kennedy Space Center). In 1980, the plant was enlarged to 291 million cubic feet to accommodate production of the 767 aircraft. Today, major parts of the assembly functions of the 747 and 767 are also housed under this one great roof.

Fallingwater, Bear Run, Pennsylvania, USA

The domestic buildings designed by the American architect Frank Lloyd Wright are amongst the most remarkable of their kind. Of these, Fallingwater is probably the best known. The house was built between 1935 and 1937 for Edgar Kaufmann, who owned Kaufmann's Department Store in Pittsburgh. Wright's first house built of reinforced concrete, Fallingwater is constructed of slabs of ochre concrete cantilevered over a waterfall with sheets of glass forming the horizontals between the planes of concrete. It reflects Wright's "organic" use of concrete, rooting his structure in the surrounding rocks by walls of rough stonework. The house has been afflicted by technical shortcomings, requiring successive heavy repairs.

Las Vegas Hilton, Nevada, USA

The largest hotel in the world, the Las Vegas Hilton occupies a 63-acre site and has 3,174 rooms and suites. There are 14 international restaurants, a casino hung with chandeliers, a 48,000-square foot ballroom and convention space with an area of 125,000 square feet. On the roof is a remarkable 10-acre recreation space, with a 350,000-gallon heated swimming pool, six tennis courts and an 18-hole putting green, as well as facilities for table tennis, badminton and shuffle board. There are 21 lifts to whisk guests up and down, and they are served in the utmost luxury by a staff of 3,600 people.

Machu Picchu, Peru

The story of the rediscovery in 1911 by Hiram Bingham of the lost Inca city, deep in the forests that cloak parts of the Andes, is one of the most romantic of archeological finds. Of all such sites, none can rival its position, surrounded by mountains and valleys on a cyclopean scale and situated at 7,975 feet above sea level. The hillsides are so precipitous that they had to be terraced, not only for the growing of food, but also to retard erosion of the soil. The terraces had a capacity to feed several hundred, and water was brought miles by aqueducts, which were still functioning when Bingham found them.

The quality of Inca masonry in the temples and houses that make up the site is as fine as that found in Cusco, where a knife blade cannot be inserted between blocks, so perfect is the join. The history of Machu Picchu is a subject of debate.

Maracana Municipal Stadium, Rio de Janeiro, Brazil

The world's biggest, this football (soccer) stadium was completed in time for the World Cup Final between Brazil and Uruguay in July 1950. It can accommodate 155,000 people seated and another 50,000 standing on the terraces. The players are isolated from the crowd by a dry moat 5 feet deep and 7 feet wide.

The Pentagon, Arlington County, Virginia, USA

The largest office building in the world, the Pentagon is the headquarters of all three branches of the armed services in the United States. Built in 1941–43, this low, five-storey block, with five sides 921 feet long, covers 34 acres, including the courtyard, and provides 3,700,000 square feet of air-conditioned floor space—sufficient for 30,000 people to work in. Built of steel and reinforced concrete with some limestone facing, it consists of five concentric rings with 10 corridors, like the spokes of a wheel, connecting them. In addition, the complex contains a huge underground shopping concourse and a heliport.

Europe

The Alhambra, Granada, Spain

Austere and formidable, the exterior of the Alhambra belies its gracious and richly ornamented interior. The conversion of the ancient fortress of Alcazába into a palace—for almost 250 years the residence and harem of Muslim rulers in Spain—began in 1238. Moorish creative genius reached its peak during the fourteenth century, when a maze of halls, columns, arcades, shady courtyards, pools and fountains was built.

With representational art forbidden by Islam, architects and artists achieved miracles of intricate abstract and geometric designs in their glazed tiles and delicate, lacelike plasterwork. The finest example of this is probably the stunning "stalactite" decorations that appear to explode in a starburst on the cupola of the Hall of the Two Sisters.

Carcassonne, Aude, France

The site of the old city, on top of a steep, isolated hill, has been occupied continuously since the fifth century BC, and towers built by the Visigoths in AD 485 can still be seen. But Carcassonne's fame rests on its medieval fortifications, the finest in Europe. Begun by the viscounts of Carcassonne in the twelfth century, they were continued after 1247 by King Louis IX of France, who constructed the outer ramparts. His son, Philip III, added further intricate defences, including the beautiful Narbonnese gate and the Tour de Trésor.

In 1355, even the redoubtable Edward, the Black Prince, found the fortress impregnable. By the end of the seventeenth century, however, the ramparts were abandoned and fell into disrepair; they were restored in the mid-nineteenth century by the great architect Viollet-le-Duc.

Castell Coch, South Glamorgan, Wales

Designed by William Burges for the 3rd Marquess of Bute and built in the late 1870s, Castell Coch bears comparison with the creations of King Ludwig II of Bavaria: both men had atavistic ideas about architecture, and produced buildings that were indulgent anachronisms. Though nominally the restoration of a castle ruinous since the sixteenth century, Castell Coch is a sham, combining the exterior appearance of a thirteenth-century Welsh castle with an interior that is one of the most exuberant products of Victorian imaginative decoration. For example, almost every surface of the vaulted Drawing Room is elaborately decorated with subjects from nature, scenes from Aesop's Fables and Greek mythology.

It is rare for an architect to be so in sympathy with a client's vision as Burges was with the Marquess of Bute; the result, here and at Cardiff Castle where the two men also collaborated, is a pair of buildings unlike any others.

Church of Notre Dame du Haut, Ronchamp, France

One of the most unorthodox church buildings ever built is the highly individual creation of Le Corbusier at Ronchamp near Belfort, erected between 1950 and 1955. Built of reinforced concrete, the church provides an interesting silhouette from all viewpoints, the roof billowing up to an acute point with an exaggerated overhang, looking like a cushion from some angles. Deeply recessed stained-glass windows of irregular shapes and size light an interior with a seeming confusion of angles and slopes.

The Colosseum, Rome, Italy

Situated near the south-east end of the Forum, the great oval Flavian amphitheatre takes its familiar name from the huge statue of Nero that stood nearby. The venue for gladiatorial fights and contests with wild animals, it was begun by Vespasian in AD 75 and inaugurated by his son Titus in AD 80. Built of concrete faced with travertine marble, at 620 feet overall and 513 wide, it is the most imposing of all remaining Roman buildings. The outer wall, 160 feet high, has four storeys, the first three arcaded with Doric, Ionic and Corinthian orders, the top one solid with Corinthian pilasters and windows. Tiers of seats, supported by concentric corridors with vaulted ceilings, could accommodate around 45,000 people. And underneath the arena, which measures 287 by 180 feet, are storerooms and dens for animals.

Gazetteer

Escorial, near Madrid, Spain

Philip II built the Escorial in 1563–84 to commemorate the Spanish victory over the French at St Quentin in 1557. The rectangular complex, measuring 675 by 525 feet, includes the large, handsome church of San Lorenzo and a mausoleum in which all Spanish sovereigns except Alfonso XIII (d1941) are buried. There are also a monastery, palace, offices, library and college, housed in five great cloisters.

The massive, sombre buildings, designed by Juan Bautista de Toledo and Juan Herrera, are made of grey granite, and are impressive rather than beautiful. The plan of the grand but austere church is that of a Greek cross, with nave and transepts of equal length, and the monumental dome is 60 feet in diameter and 320 feet high. Today the Escorial contains a magnificent collection of paintings, rare books and manuscripts.

Fonthill Abbey, Wiltshire, England

The now-vanished country house at Fonthill was one of the most fantastic houses ever built in a country not short of eccentric creations. Its creator, William Beckford, had inherited a huge fortune, derived largely from plantations in the West Indies. An admirer of the Gothic, Beckford commissioned James Wyatt to design a pile that would stand comparison with nearby Salisbury Cathedral. Eleven years of round-the-clock work by two teams, each of 500 workmen, were required to build the cruciform structure, completed in 1808.

The main part measured 312 feet by 250 feet, the Great Hall's ceiling was 80 feet high, and there were two Long Galleries, but it was the tower surmounting the crossing that astonished the few visitors admitted by the reclusive Beckford: the octagonal spire soared 276 feet into the air, but it was to prove the building's undoing. Beckford's impatience to see his house completed and a lack of scruples by the builder combined to encourage the latter to skimp on the foundations. Crucial inverted arches had been omitted, with the result that the tower fell in 1825, though not before Beckford had sold the Abbey. It was never rebuilt, and within thirty years of its construction the rest of this vast building had disappeared.

Hagia Sophia, Istanbul, Turkey

A pool of serenity in the maelstrom of modern Istanbul (formerly Constantinople), the great Byzantine church of Hagia Sophia, "Holy Wisdom", was built by Justinian in 532–37. The third church to stand on the site, the exterior is a jumble of semi-domes and buttresses, with four minarets at the corners, added later. But the interior, with an area of 9,800 square yards, and the dome, with a diameter of more than 100 feet, are magnificent.

Justinian imported red porphyry, verdantique, and yellow and white marble, and employed sculptors and mosaic artists to create the finest church in Christendom. When Constantinople fell to the Ottoman Turks in 1453, Hagia Sophia was converted to a mosque and the mosaics plastered over. Finally, in 1934, it became a museum.

Herrenchiemsee and Neuschwanstein, Bavaria, West Germany

Several of the world's most opulent and fantastic buildings were built by the eccentric, romantic King Ludwig II of Bavaria, patron of Wagner.

Herrenchiemsee, built on the largest of three islands in Bavaria's largest lake, was Ludwig's Versailles. The foundation stone was laid in 1878, and by the time of Ludwig's death in 1886, only the central block and part of one wing had been finished. But it already contained some of the most magnificent objects made for any palace: the largest porcelain candelabrum in the world, produced by Meissen; the Hall of Mirrors that eclipses that at Versailles; curtains that each weigh a hundredweight; a door with Meissen plaques in the panels.

In contrast, Neuschwanstein is a pastiche medieval castle, built on a spectacular mountain top of which over 20 feet had to be blasted away to provide a level surface. It is the castle's dramatic setting, ringed by Alpine peaks, that makes this fairytale fantasy so memorable. Work began in 1869 and the finishing touches were still being done at Ludwig's death. Besides the traditional rooms of the castle, the king's study gave access to an artificial grotto with cascade and variable lighting effects to match the monarch's mood.

Knole, Kent, England

Seen from a distance, across its 1,000-acre park, the country house of Knole resembles a medieval town. It is reputed to be the house (as opposed to palace) with the largest number of rooms—365. Eventually built around seven courtyards, the house was begun in 1456 by the Archbishop of Canterbury Thomas Bourchier, who bequeathed it to the see of Canterbury. Henry VIII coerced Archbishop Cranmer into giving it to him, and the acquisitive king enlarged it considerably, almost certainly adding the imposing west front, 340 feet long, for the enormous retinue that accompanied the monarch and visiting ministers. The Great Hall, originally intended as the room in which the household ate, measures 95 feet by 32 feet. The Cartoon Gallery is even longer, at 135 feet.

The Leaning Tower, Pisa, Italy

This round Romanesque tower, the campanile to the nearby Baptistery, was started in 1174 by Bonanno Pisano. Built entirely in white marble, with eight tiers of arched arcades, the tower is 179 feet high. It began to lean as soon as the first storey was completed, probably because the alluvial subsoil settled or the foundations were inadequate. Ingenious attempts were made to compensate for the tilt by straightening up subsequent storeys and making the pillars higher on the south side than the north. The bell chamber, finished only in 1350, was also built at an angle and the heaviest bells hung on the north, but the tower continued to lean and is currently some 17 feet out of perpendicular.

Lincoln Cathedral, England

Regarded as the finest example of Early English architecture, Lincoln Cathedral had the distinction of being the world's tallest structure between 1307, when its central spire reached 525 feet (overtaking the Pyramid of Cheops in Egypt), and 1548 when it fell in a storm. The traveller and writer William Cobbett even thought it "the finest building in the whole world". Its position atop the hill that dominates Lincoln gives it a commanding site unrivalled by any cathedral but Durham.

Building began c1075 under Bishop

Remigius and the cathedral was dedicated in 1092. Damage sustained through an earthquake in the twelfth century necessitated its rebuilding under the bishop St Hugh of Lincoln, begun *c*1190. Amongst the glories of the cathedral are hundreds of statues that adorn the exterior, the quality of the carving on the choir screen and in the Angel Choir, and the library designed by Christopher Wren. The spires that once crowned the two shorter, western towers were removed in the eighteenth century, despite the riotous protest of the townspeople.

Nat-West Tower, Old Broad Street, London, England

The tallest cantilevered building in the world, and the tallest office block in Britain, is the National Westminster Bank's 600-foot tower in the City of London. It has three basement levels and 49 storeys, which rest on steel and concrete supports projecting from a central tower. It was designed by Richard Seifert and completed in 1979.

The Palm House, Kew Gardens, London, England

After Queen Victoria's visit to Paxton's lily-house at Chatsworth (see p.64), it was suggested that the Royal Botanical Gardens at Kew should build a comparable structure. The two architects involved were Decimus Burton and Richard Turner: Burton had helped Paxton with the Chatsworth building, and Turner had contributed to the Palm House for the Botanic Gardens in Belfast. Work began in 1844 on the designs for what was to be the longest such building, 362 feet long, 100 feet wide in the centre and 63 feet high. Cast-iron columns supported the curved mullions of the walls and roof. A separate boiler house provided heat through pipes buried in a walk-through tunnel.

The Parthenon, Athens, Greece

The Parthenon was built as a temple to the goddess Athena in 447–438 BC. In essence, it consists of a rectangular base, 238 by 101 feet overall, with a colonnade on all four sides enclosing the two small rectangular chambers of the naos—the city's treasury and a room to house the sumptuous gold and ivory statue of Athena. The roof was low pitched, with a triangular pediment at each end. But these geometric forms are softened and enlivened by subtle variations that make the Parthenon the most perfect of ancient Greek buildings. All horizontal lines curve upward toward the centre, and the columns, which swell slightly in the middle and taper toward the top, lean inward.

The temple is built of marble blocks, precisely fitted without mortar and decorated with carvings, embellished with bronze and gold. A carved frieze ran around the top of the naos. Carved panels above the pillars and magnificent high-relief carvings on the pediments depicted the birth of Athena and the battle between Athena and Poseidon, the sea god, for Athens.

Petra, Jordan

In 1812 the Swiss traveller J.L. Burckhardt rediscovered the ancient city of Petra, "a rose-red city half as old as time", which flourished as the Nabatean capital for about 500 years from the second century BC to the early fourth century AD. What make Petra one of the world's most spectacular archeological sites are its setting, ringed by barren mountains, the fabulous carving work of the Nabateans, and the overlay of Roman buildings that was erected after their annexation of the city in AD 106. Petra is reached by a narrow gorge of ½ mile, the Siq, that runs between almost vertical cliffs 300 feet high. This approach was impregnable, capable of being defended by a handful of soldiers.

The Nabateans were amongst the world's finest carvers of stone and the El-Khazneh, or Treasury, which is the first building seen at the end of the Siq, is the most spectacular example. Carved out of orange-pink rock, the Greek-style building was probably a temple. The Nabatean buildings that were carved into the cliff faces have been protected by the overhang and are much better preserved than the later Roman buildings. Of these, the best preserved is the amphitheatre which once held over 3,000 spectators in 33 tiers of seats. Three markets, temples, a forum, baths, gymnasia, colonnades, many shops and private houses once covered the Roman site, 2 miles long.

Pompeii, Naples, Italy

Founded in the fifth century BC, Pompeii came first under Greek influence; but by AD 79, when it was overwhelmed by the eruption of Vesuvius, it had become a town of 25,000 and a summer resort for wealthy Romans. Systematic excavation of the ruins began only in 1748, and a third of the city still remains buried.

The villas, temples, baths, civic buildings, forum and amphitheatre so far uncovered are mainly in the Roman style. They are built of brick, faced with marble or plaster, and some, notably the Casa dei Vettii, are richly decorated with frescoes. The paved streets, between high pavements, are deeply rutted by chariot wheels and frequently crossed by stepping stones for pedestrian use. The remains of buildings and people as they emerge provide graphic and moving evidence of the life of this ancient city.

Pompidou Centre, Paris, France

Designed by Richard Rogers and Renzo Piano, this huge museum and display centre, completed in 1975, is one of the most controversial of modern buildings. It has no formal façade and is constructed of gigantic steel beams and trusses painted in bright primary colours— red, blue and yellow—which are clearly visible through the glass walls, as are the external escalators and connecting galleries.

The structure consists of five floors some 360 feet long and 160 feet wide, with a public forum 1,368 square yards in extent on the ground floor. It also houses a museum of modern art with 44,800 square yards of exhibition space, research institutes for industrial creation and for music and acoustics, a library and a restaurant.

Ponte Vecchio, River Arno, Florence, Italy

Built by Taddeo Gaddi in 1345, the Ponte Vecchio was the first bridge in the West with arches smaller than a semicircle. This meant that fewer piers were needed to support it, affording a freer passage to boats and, most importantly, to flood water, for the Arno carries vast quantities of melt-water in spring.

On either side of the roadway is a two-

Gazetteer

storey gallery. The top storey acts as a corridor, linking the Uffizi Palace, which housed the offices of the ruling Medici family, with the Pitti Palace on the other side of the river. At street level there were, and still are, shops occupied by gold- and silver-smiths and jewellers.

Prague Castle, Hradčany, Prague, Czechoslovakia

Founded *c*850 as a wooden keep, the "castle" at Hradčany, the largest in the world, is, like the Kremlin in Moscow, now a complex of buildings rather than a fortress. Grouped around three courtyards, the irregular oblong of buildings covers 18 acres. A visitor penetrating the castle enters through the elaborate Matthias Gate, then passes from early twentieth-century architecture back through baroque, renaissance and Gothic to the huge medieval White and Dalibarka towers.

The most impressive buildings are the Royal Palace and the Gothic cathedral of St Vitus. Designed by Matthias of Arras in 1344, it is the third on the site; the original was founded *c*930 by Prince Wenceslas, now the country's patron saint. From at least 894, the castle has been the official seat and coronation place of Czech sovereigns; even today, the president is inaugurated in the palace in the vast, late Gothic Valdislav Hall, which is 243 feet long, 60 feet wide and 50 feet high.

The Royal Pavilion, Brighton, Sussex, England

Originally a small farmhouse, the Royal Pavilion was changed and added to over 35 years, from 1786 to 1821, to satisfy the whim of the Prince of Wales, later King George IV. Its final fantastic pinnacles and domes, owing much to Indian Islamic architecture, date from 1815. They were the work of John Nash, who for the first time in domestic architecture made much use of cast iron – other than for window frames and fireplaces – both structurally and decoratively.

Astonishing as the exterior is, the interior is even more stunning, for Nash added the huge domed banqueting hall and music room and decorated both with wild extravagance. Great dragons curl around the music room's walls and deep-red and blue ceiling, encrusted with gold; and in the exotically painted banqueting hall hangs a jewelled chandelier weighing almost a ton.

St Paul's Cathedral, Ludgate Hill, London, England

The first stone of Christopher Wren's masterpiece was laid in 1675 and the cathedral was finished in 1710. The three-aisled nave and choir are 463 feet long and 101 feet wide, and over the crossing is a shallow dome, 450 feet in diameter and 218 feet high, covered in mosaics. The dome is supported by pillars, grouped at the corners to house church offices and a staircase to the library. These pillars were originally built of stone, infilled with rubble; in the 1930s they were strengthened by the injection of liquid concrete, which perhaps helped them to withstand destruction in the bombing during World War II.

The dome consists of three shells. Above the inner one is a conical brick structure which carries the timber framework of the external lead-covered dome. It also supports the colonnaded lantern and a great golden cross that rise to 365 feet above the ground.

On the west front are two 222-foot towers, in the southern one of which hangs the 17-ton bell, Great Paul, the largest in England.

The Palace of Versailles, Versailles, France

Built on the site of a royal hunting lodge, the palace of Versailles was the triumphant product of the ambition of Louis XIV and the designs of the great architects in the classical style: Le Vau, Le Brun and Hardouin-Mansart. Begun in 1661, it was 50 years in the building. The vast palace and gardens occupy 6,000 acres.

The west, garden front, 2,197 feet long, was built largely by Le Vau in 1669, and in 1678 its open terrace was enclosed by Mansart to create the most splendid room in the palace, the Hall of Mirrors, 238 feet long. Here, 17 tall arched windows are matched by 17 dummy arches lined with mirrors set in gilded copper, and both windows and mirrors are separated by red marble pillars. A gilded stucco cornice frames the painted ceiling. The room was originally furnished with silver furniture and chandeliers, and sumptuous carpets, to reflect the Sun King's magnificence.

Windsor Castle, Berkshire, England

Begun by William the Conqueror in 1067, the royal residence of Windsor Castle is the largest inhabited castle in the world. Its plan is roughly a figure of eight, and the curtain wall extends for over ½ mile. The massive cylindrical shell keep, which dominates the town's skyline, was built by Henry I, who first used stone in the castle's construction. The keep's height has since been increased to 100 feet and the exterior wall refaced. By the end of Edward III's reign, the castle had ceased to be primarily a military stronghold and become principally a residence. Successive monarchs have altered and extended the castle, but its present appearance resembles its medieval form. Foremost of the additions is the Garter chapel of St George, begun by Edward IV.

Rest of the World

Angkor Wat, Angkor, Kampuchea

The temple of Angkor Wat, one of the largest religious complexes in the world, covers an area of nearly 1 square mile. Built of sandstone in 1113–50 under the ruler Suryavarnam II as his sepulchre, it was dedicated to the god Vishnu and represents the Hindu cosmology.

The great temple, surrounded by a moat representing the oceans, is approached across a causeway 1,000 feet long. A magnificent gateway in the outer wall—the mountains at the edge of the world—gives access to five concentric rectangular enclosures, overlooked by towers in the form of lotus buds, the tallest more than 200 feet high. The five central towers denote Mount Meru's peaks, the hub of the Universe. The courts are linked by colonnades lined with elaborate sculptures and immense decorative bas-reliefs depicting scenes from sacred Hindu legends—the temple's most remarkable feature.

Chandigarh, Punjab, India

Le Corbusier's vision for an ideal city received partial expression in the administrative capital of the Punjab, founded in 1951, for which he designed the principal buildings: the Palace of the Assembly (parliament), High Court and Secretariat.

Their massive dimensions make them appropriate symbols of government, but they have proved less than functional, not least because of the space between them—uncomfortable in the heat of the Punjab. The radical departure from any national architectural tradition was a deliberate policy of Prime Minister Nehru at the beginning of India's independence, intending the buildings to be "symbolic of the freedom of India".

Fatehpur Sikri, Uttar Pradesh, India

In 1569, the emperor Akbar built a mosque and tomb at Fatehpur Sikri to honour the hermit Salim Chisti, who had foretold the birth of his son, later the emperor Jahangir. Public buildings and palaces were the next to be erected, and the city, encircled by battlemented walls, became Akbar's capital until 1588. By 1605, due to a shortage of water, it was deserted.

Built of soft, rose-coloured sandstone, easily carved, Fatehpur Sikri is an exquisite and almost intact example of Moghul architecture. Most remarkable are the Buland Darwaza (victory gate) with its immense elephant statues; the fine façade of the Jami Masjid (great mosque); Salim's marble mausoleum with its fine traceries and enamel and mother of pearl inlays; and the palaces of Jodh Bai and Birbal.

Great Zimbabwe, Zimbabwe

The largest stone monument in Africa outside Egypt is the complex of ruins 250 miles inland from the Indian Ocean port of Sofala, known as Great Zimbabwe. The buildings of the Acropolis, or hill fortress, look down on those in the valley that lie within the Great Enclosure. This dry-stone wall, 830 feet in circumference, 16–35 feet high and at least 4 feet thick at the base, is made from blue-grey granite cut and laid like bricks. Inside are other walls forming narrow passages, three platforms, several "chambers" and a solid stone conical tower.

The builders of Great Zimbabwe and its purpose are not known, but it was probably founded around the tenth century as a centre for trade in artefacts from a flourishing Iron Age community and in black slaves for Arabia.

Hong Kong and Shanghai Banking Corporation Headquarters, Hong Kong

This masterpiece of technical innovation, designed by Norman Foster and completed in 1986, is made up of three visually distinct towers of different heights, the 47-storey central section rising to 590 feet. The entire building is suspended over an open ground-floor plaza from eight immense steel towers clad in aluminium panels. The exposed steel structure is divided into five vertical zones within which, in a method derived from bridge construction, a stack of lightweight steel and concrete decks is hung from steel suspension trusses that look like giant coathangers. These stacks were built largely from prefabricated modules made in USA, Britain and Japan.

On the south face of the building is a computer-controlled "sunscoop" of 24 mirrors, which follows the sun and reflects its rays on to the top of the 150-foot high central atrium; from here it floods down throughout the building.

Hoysaleswara Temple, Halebid, Karnataka (Mysore), India

The Hoysala dynasty, which ruled in this region for some 250 years until 1326, reached the peak of its power under Bittiga (1110–52), who took the name of Vishnuvardhana when he was converted to Hinduism. The most remarkable of the temples he erected to honour his new religion was the Hoysaleswara temple at Halebid, his capital city.

In itself, the fairly small, squat, star-shaped building is not impressive; it is the intricate sculptures with which every surface is covered that establishes the temple as the acme of Hoysala artistic achievement. Carved in soft soapstone that hardens with exposure are episodes from the lives of the princes: hunting scenes, depictions of rural life, animals, birds, and, above all, musicians and dancing girls.

There is also a huge statue of the Jain god Gommateshwara and of the bull of the Hindu god Shiva.

New Delhi, India

The city of New Delhi, on the right bank of the Jumna River, was designed by Edwin Lutyens and Herbert Baker. It was built between 1912 and 1929 to replace Calcutta as the capital and administrative centre of British India. Its broad streets are symmetrically laid out to afford wide views of fine government buildings and historical monuments, including a huge war memorial arch erected in 1921. From this arch, a broad, tree-lined avenue, the Raj Path, leads to a magnificent marble and sandstone palace. Originally the Viceroy's Palace, since Independence it has become the official residence of the Indian president.

In a prayer ground in the south of the city, Mahatma Gandhi was assassinated in 1948.

Polonnaruwa, near Sigiriya, Sri Lanka

Built in a wonderful site beside a lake, the ancient city of Polonnaruwa was once the most magnificent in Sri Lanka, formerly Ceylon. It became a royal residence as early as 368 and during the eighth century was the capital of the island. Its period of greatest importance was during the reign of the most famous Singhalese king, Parakrama Bahu I, who reigned from 1164 to 1197, and the principal ruins date from this time. The most imposing is the Jetawanarama temple, 170 feet long, whose walls reach 80 feet in height and are 12 feet thick, and the immense reclining statue of Buddha.

The Potala, Lhasa, Tibet

Its thousand windows and gleaming golden roofs visible from miles away, the imposing and powerful structure of the Potala rises high on a hill above Lhasa. For more than 300 years, until the Chinese annexation in 1951, it was the fortress-palace of the Dalai Lamas, Tibet's spiritual leaders; today it is a museum.

The whitewashed stone walls of the outer White Palace, completed in 1648, enclose the Red Palace, finished in 1694. This is the religious centre of the complex, with the monks' assembly halls, libraries of Buddhist scriptures, chapels, shrines and, most impressive, the 50-foot-high funerary pagoda of the Fifth Dalai Lama—the Potala's founder. It is made from sandalwood and covered with 4 tons of gold, studded with diamonds, rubies and sapphires.

Gazetteer

Taj Mahal, Agra, India

One of the best-known buildings in the world was an extraordinary indulgence, a personal celebration of the love felt by the seventeenth-century Moghul emperor, Shah Jahan, for his queen, Mumtaz Mahal. She died in 1631 after bearing 14 children during their 17 years of marriage. Work on the building began in the same year.

For the next 20 years, 20,000 men and women toiled to turn the drawings of an architect, whose identity remains a mystery, into the gleaming white mausoleum: skilled craftsmen were recruited from all over Asia; elephants and bullocks dragged countless blocks of marble along a 10-mile ramp of tamped earth to the construction site. The surfaces of the Taj were inlaid with precious and semi-precious stones until the troubled eighteenth century, when they were stolen. The dome is the most imposing part of the building, the finial reaching 220 feet.

The Taj was neglected after Shah Jahan's sons died, and under the Raj there was even a plan to dismantle it and sell the marble in England. Only with the revived interest in India's architectural heritage, fostered by Lord Curzon, were the mausoleum and its grounds renovated.

Schwedagon Pagoda, Rangoon, Burma

In a "nest of hundreds of smaller pagodas", on a 14-acre hilltop site dominating the city of Rangoon, is the Schwedagon Pagoda, the most magnificent Buddhist shrine in Burma.

Legend has it that the first pagoda was built here in the sixth century BC; the present stupa, dating from 1768, was raised by King Hsibyushin to replace one destroyed in an earthquake. The bell-shaped central body, which stands on a series of rectangular and octagonal terraces—the whole plated with pure gold—rises more than 300 feet in tapering sections to a gilt-iron *hti*, or "umbrella", hung with gold and silver bells. This is surmounted by a gem-encrusted vane and an orb studded with 4,000 diamonds.

FEATS OF CIVIL ENGINEERING

The Americas

Chesapeake Bay Bridge-Tunnel, Virginia, USA

The world's longest bridge-tunnel was opened to traffic in April 1964 after just 42 months' work and the expenditure of $200 million. The $17\frac{1}{2}$-mile combination of trestles, bridges and tunnels links Norfolk and the tip of Cape Charles. To keep the channel into Chesapeake Bay clear for shipping, two concrete-lined tunnels a mile long and 24 feet in diameter were taken deep beneath the main channel, joining two man-made islands each 1,500 feet long. The main part of the crossing is on $12\frac{1}{2}$ miles of precast concrete pile trestles, 31 feet wide and 25 feet above mean low water, and capable of withstanding 20-foot waves.

Golden Gate Bridge, San Francisco, USA

Although over 50 years old, this bridge is still regarded as one of the world's great civil engineering masterpieces. When Joseph Strauss completed his design for the bridge in 1930, it was for the world's longest span, at 4,200 feet, overtaken only in 1964 by the Verrazano Narrows Bridge, New York, which has a span of 4,260 feet.

More than 100,000 tons of steel, 693,000 cubic yards of concrete and 80,000 miles of cable were used in its construction. The overall length, including freeway approaches, is 7 miles. The towers are 746 feet high and there are two supporting piers, the larger of which extends 100 feet below the sea. At low tide, the roadway is 220 feet above the water.

The biggest obstacle encountered by the engineers on the Golden Gate Bridge was the construction of the foundations on account of the strong tides; deep-sea divers could work for just four periods of 20 minutes a day when the tide turned and the water was relatively slack. Several abortive attempts were made to establish a caisson within which to build the piers of the bridge, compelling the engineers to build instead a cofferdam—a watertight case

kept dry by pumping. The bridge opened in May 1937 and had cost $35 million.

Quebec Bridge, Canada

The world's longest cantilever truss span, at 1,800 feet, is situated some way up the St Lawrence from Quebec to take advantage of a narrowing of the river from its usual width of 2–3 miles to little more than $\frac{1}{4}$ mile. An earlier bridge at this point collapsed, and construction of the present bridge was begun in 1899, but was fraught with problems. Warnings of excessive deflection as work progressed were ignored until, in 1907, the bridge collapsed, killing 75 workmen. An inadequate number of rivets in one of the cantilever arms and the buckling of a web member were the cause.

The new design was for a much stronger bridge, 3,300 feet long and entailing the use of 150 percent more steel. However, disaster struck again, in 1916, when the suspended span was being lifted into place and a casting broke; the 640-foot, 5,000-ton span fell into the river, killing 13 workers. The third attempt was successful, and the opening train steamed across in December 1917. A roadway was added in 1929.

Niagara Suspension Bridge, Niagara River, USA and Canada

Amid dire predictions that it would collapse in the first high winds, the first modern suspension bridge was opened to rail and passenger traffic at Niagara in 1855. That the double-decker bridge, with a main span of 821 feet, proved safe was due to the recognition by its designer, John A. Roebling, that a suspension bridge needed to be not only strong but stable. Strength was gained from two $10\frac{1}{4}$-inch diameter cables on either side, each of which supported one of the decks, each 10 feet wide. Stability was ensured by the 64 stays and deep timber trusses between the decks. But the iron cables gradually deteriorated, and in 1897 Roebling's suspension bridge was replaced by a steel arch structure, itself replaced by the Rainbow Bridge.

Second Lake Washington Bridge, Seattle, Washington, USA

The longest floating bridge in the world—12,596 feet overall with a 7,518-foot floating section—is the Lacey V. Murrow Bridge, finished in 1963, which traverses Lake Washington on Interstate 90. The lake was too deep, at 150 feet, to bridge conventionally, but there are no currents and no ice, so a pontoon bridge was the ideal solution. Each of the 25 reinforced concrete pontoons measures 350 feet long, 60 feet wide and 14 feet to the roadway; pontoons are divided internally into watertight compartments. As well as the floating section, there are three reinforced concrete girder spans to allow small ships to pass.

Europe

Southend Pier, Essex, England

The world's largest pier was built to serve a resort that was already popular by the beginning of the nineteenth century. The first, wooden, pier was begun in 1829 and was extended from 1,800 feet to 1¼ miles in 1846. In 1887 it was almost entirely rebuilt by Arrol Bros to a design by J. Brunlees. The new pier was 6,600 feet long, and was improved and extended on several occasions, to reach almost 7,000 feet.

Piers had two principal functions: to enable visitors to walk away from the beach into healthier sea air, and to enable pleasure steamers to ply the seaside resorts without the use of rowing boats to ferry passengers ashore. Many piers incorporated a theatre or at least a café and shops. At Southend an electric narrow gauge railway linked the shore with the three-tiered pavilion structure at the seaward end where coastguard and lifeboat stations were provided.

Forth Railway Bridge, Fife, Scotland

The graceful design for a cantilever railway bridge across the Firth of Forth by John Fowler and Benjamin Baker (who designed the tube for the journey of Cleopatra's Needle to London) was the first use of steel for a major bridge in Europe. Suspicion over the use of steel after disappointing results in some Dutch railway bridges led to a Board of Trade prohibition on the metal's use for bridges until 1877. The three piers support cantilever arms of 680 feet which are joined by two suspended spans, each of 350 feet, making two main spans of 1,710 feet, which made it the world's longest span between its opening in 1889 and completion of the Quebec Bridge in 1917.

Royal Albert Bridge, Saltash, Devon, England

The last great work by the brilliant and versatile engineer Isambard Kingdom Brunel was the viaduct taking the Cornwall Railway across the Tamar estuary into Devon. It was notable in being the first major work in which compressed air was used to expel water from a caisson—a working chamber in which foundations can be dug below the surrounding water.

Brunel developed the principle behind his earlier, smaller bridge at Chepstow to produce a twin-span bridge reached by short, curving approach spans. The two central spans are a combination of arch and suspension bridge, the upper chord consisting of a huge wrought-iron oval cylinder from which two chains are suspended to form the lower chord. The spans, each of 455 feet and weighing 1,060 tons, were built on the river bank and floated into position. Brunel did not witness the opening by the Prince Consort in May 1859, being critically ill; he died four months later.

Pontcysyllte Aqueduct, Shropshire, England

Thomas Telford's aqueduct across the Dee Valley broke new ground in using for the canal trough and towpath a material new to this application—cast iron. By the time construction began in 1795, a number of cast-iron arch bridges had been built, following Abraham Darby's pioneer structure of 1779 at Ironbridge. But Telford incorporated the canal trough itself in wedge-shaped sections, which bolted together through flanges at each end, the wedges building up to form arches that rested on masonry piers. Even after building an embankment at each end, 19 spans each of 53 feet were necessary to bridge the valley, making the aqueduct 1,007 feet in length. Opened in 1805, the aqueduct still carries boats on the Ellesmere Canal.

Afsluitdijk, Zuider Zee, Netherlands

The Zuider Zee was formed in the thirteenth century when the North Sea, surging inland, captured an existing lake. Over the centuries, attempts were made to reclaim the flooded land, but large-scale reclamation only became possible through one of the world's greatest feats of civil engineering, the construction in 1927–32 of the Afsluitdijk, a sea dam 20¼ miles long and 25 feet high. Built in two parts, this split the Zuider Zee into the Ijsselmeer and the Waddenzee.

In shallow areas, two walls of boulder clay were built up and sand pumped into the space between them; the sloping dam walls were faced with brushwood bundles and stone. In two deep areas, sill dams extending 11½ feet below the average sea level were built first. The dam's width at sea level is 293 feet.

Four polders—reclaimed land—totalling 700 square miles have already been returned to agricultural and urban use, and another 155 square miles will be added with the completion of a fifth polder, the Markerwaard. The Ijsselmeer will become a freshwater lake 540 square miles in extent.

Thames Barrier, River Thames, Woolwich, London

The barrier across the Thames—1,706 feet long and 105 feet high—is the world's largest tidal river barrier. After almost 13 years of planning and building, it was opened by Queen Elizabeth II in 1984. Designed to protect vulnerable areas along the river from flooding by the North Sea, it consists of 10 movable steel gates, 9 piers and 2 abutments. Four large rising sector gates, each 200 feet wide and weighing some 1,300 tons, span the main navigational channels; two 103-foot rising sector gates provide two narrower channels; and adjacent to the abutments are four falling radial gates.

To facilitate shipping, when the rising sector gates are opened, they rotate about 90 degrees until their curved surface is housed in a shaped recess on the river bed and their flat upper surface is flush with the river bottom. The gates are operated by a hydraulically powered mechanical system.

Tower Bridge, River Thames, London

Probably the best-known sight in London is the elaborate gothick Tower Bridge over the Thames. The first bridge encountered by shipping, it was designed by Horace Jones and John Wolfe Barry and built between 1886 and 1894. The bridge is made of iron, clad in Portland stone and grey granite, and consists of two counterweighted bascules (from the French for seesaw) and two suspended spans that connect the 205–foot main towers with the shore. Each of the bascules is made up of four main 100-foot girders, with cross-bracings, and weighs around 1,000 tons. Originally they were operated hydraulically and took about 6 minutes to open; today they are opened electrically in $1\frac{1}{2}$ minutes. Pedestrians can cross the bridge even when it is open by means of high-level walkways between the towers.

Rest of the World

Carthage Aqueduct, Tunisia, North Africa

The longest aqueduct of ancient times was this channel of $87\frac{1}{2}$ miles, built by the Romans during the reign of Emperor Hadrian (117–138). It brought water from the inland springs of Zaghouan to huge underground cisterns built earlier by the Carthaginians at Maalaka on the outskirts of their city.

The piers supporting the channel stood some 15 feet apart and were 15 feet high by 12 feet thick. The channel itself was 3 feet wide and 6 feet high, and it has been calculated that it had a capacity of 7 million gallons a day.

"The Giant Peter", Himeji Central Park, Hyogo, Japan

The largest Ferris wheels in the world today are "The Giant Peter" and its companion wheel at Tsukuba, also in Japan, which can carry 384 people. Both wheels measure just under 279 feet in diameter—29 feet more than the original wheel designed by George Ferris, which was built at Chicago in 1893.

Kariba Dam, Zambezi River, Zambia and Zimbabwe

Downstream of the Victoria Falls, where the great Zambezi River used to thunder through the Kariba Gorge, stands the Kariba Dam. One of the world's largest dams and, at 420 feet, the fourth highest in Africa, the concrete arch structure measures 1,900 feet along the crest. The dam was built in 1955–59 and first filled in 1963. Some 50,000 people living along the banks of the Zambezi had to be resettled as Kariba Lake, the reservoir formed by the dam, grew to 175 miles long and 32 miles across at its widest, and thousands of wild animals had to be moved to safety. Kariba's hydroelectric project supplies almost all Zambia's electricity needs, and also serves a large part of Zimbabwe.

Mohammadieh Noria, Hamah, Syria

Where river banks are steep, as on the River Asi (Orontes) at Hamah, one of the most efficient ways of raising water is by means of a noria, or waterwheel. This wheel, one of several at Hamah dating from Roman times, has a diameter of 131 feet and is the largest in the world. The tall, undershot wheel is lightly built from timber in a complex design, with a chain of scoops around the rim. As the revolving scoops reach the river, they fill, and at the top discharge the water into a towering aqueduct which carries it to the fields for irrigation.

Nurek Dam, Vakhsh River, Tadjikistan, USSR

The highest dam in the world is this earth embankment dam with a clay core close to the border with Afghanistan. Started in 1962 and finished only in 1980, the dam is 984 feet high and 2,310 feet along the crest and was designed to withstand severe earthquakes, which are common in the region. Water from the dam, which has a volume of 2,048 million cubic feet, is used to generate electricity and to irrigate more than $2\frac{1}{2}$ million acres of land in the Amu-Darya area.

UNDERGROUND ENGINEERING ACHIEVEMENTS

Europe

Thames Tunnel, London, England

The tunnel that links Wapping and Rotherhithe is arguably the most historic tunnel ever built: it was the first underwater tunnel and the first to be built with the use of a tunnelling shield, which became the normal method of tunnel boring. The shield's function is to protect the roof and sides of the tunnel until the lining of brickwork has been completed, and to facilitate manual or mechanical excavation. The credit for these distinctions goes to Marc Brunel and his even more remarkable son, Isambard Kingdom Brunel, who at the age of 20 was engineer in charge.

The Brunels began work in March 1825, sinking a shaft at Rotherhithe in which the tunnelling shield was installed. Work progressed more slowly than anticipated, partly due to the difficult nature of the ground, which also made working conditions in the tunnel extremely injurious to health. Two inundations flooded the workings, the second almost killing the younger Brunel and causing work to come to an end through lack of finance. A government loan enabled work to be resumed after seven years. By this time, Marc Brunel had improved the design of his tunnelling shield, and he drove the tunnel through to Wapping by 1843. It became a tourist attraction, venue for art exhibitions and markets, as well as a way for pedestrians to cross the river. Its cost and the modest returns compelled the company to sell out in 1865 to the East London Railway, which adapted it for steam trains. Today electric Underground trains still use the twin tunnels.

Metro, Moscow, USSR

Moscow has the third largest metro system in the world, at 132 miles. The first section, constructed by the cut-and-cover method, was opened in 1935. Work was carried out largely by pick and shovel under the direction of the future Soviet leader Nikita Krushchev. During World War II, the 16 miles of completed line acted as an air raid shelter, as did the London Underground tunnels. Construction continued during the war, the first deep-level tube opening in 1943. Subsequent lines have been built at a depth of 100–160 feet, which is deeper even than the London deep-level tubes.

Moscow's Metro is famous for the opulence of some of its stations and for their sense of spaciousness; some are decorated with marble, ornate plasterwork, chandeliers and murals. Despite being the busiest system in the world, with 2,500 million passenger journeys a year, it is probably also the cleanest: litter or graffiti are rarely seen.

ASTRONOMICAL CONSTRUCTIONS

Europe

William Herschel Telescope, La Palma, Canary Islands

Sited above the clouds, almost 8,000 feet up on the volcanic island of La Palma, is the world's third largest single mirror telescope, and the most powerful. Named after the eighteenth-century astronomer, the altazimuth telescope, driven by minicomputers and pointed and supported by 10,000 tons of equipment, took 12 years to build. It was completed in 1987.

The 17-ton mirror, $14\frac{1}{2}$ feet in diameter, is made of special non-expanding glass ceramic, polished to an accuracy of 10,000th of a millimetre and then covered with a film of aluminium weighing 1/57th of an ounce. The telescope, so sensitive that it could detect a candle flame 100,000 miles away, is used to collect photons—particles of light—which are deflected on to a mass of detection devices to give astronomers information about immensely distant heavenly bodies.

The most important of these intstruments is the spectrograph, which splits light into its constituent rainbow colours; from the shift of the colour band toward one or other end of the spectrum, astronomers can tell if the star is approaching the earth or moving away from it.

Bibliography

Beaver, Patrick. *A History of Tunnels* Peter Davies, 1972.

Beaver, Patrick. *The Crystal Palace* Hugh Evelyn, 1977.

Bergere, Thea and Richard. *The Story of St Peter's* Frederick Muller, 1966.

Boyd, Alastair. *The Essence of Catalonia* André Deutsch, 1988.

Briggs, Asa. *Iron Bridge to Crystal Palace* Thames and Hudson, 1979.

Coe, Michael D. *Mexico* Thames and Hudson, 1962.

Cossons, Neil and Barrie Trinder. *The Iron Bridge* Moonraker Press, 1977.

David, A. Rosalie. *The Egyptian Kingdoms* Elsevier, 1975.

Descharnes, Robert and Clovis Prevost. *Gaudi* Bracken Books, 1971.

Dmitrev-Mamonov, A.I. and A.F. Zdiarski. *Guide to the Great Siberian Railway* David & Charles Reprints, 1971.

Dorin, Patrick C. *Canadian Pacific Railway* Superior Publishing, 1974.

Fedden, Robin and John Thomson. *Crusader Castles* John Murray, 1957.

Fryer, Jonathan. *The Great Wall of China* New English Library, 1975.

Garlinski, Jozef. *Hitler's Last Weapons* Julian Friedman, 1978.

Ghosh, A. (Ed.). *Jaina Art and Architecture* Bharatiya Jnanpith, 1974.

Gladstone Bratton, F. *A History of Egyptian Archaeology* Robert Hale, 1967.

Gorringe, Henry H. *Egyptian Obelisks* C.H. Yorston, 1885.

Habachi, Labib. *The Obelisks of Egypt* Dent, 1977.

Hayden, Richard Seth and Thierry W. Despont. *Restoring the Statue of Liberty* McGraw-Hill, 1986.

Hayward, R. *Cleopatra's Needles* Moorland, 1978.

Hopkins, H.J. *A Span of Bridges* David & Charles, 1970.

Hughes, Quentin. *Military Architecture* Hugh Evelyn, 1974.

Hunter, C. Bruce. *A Guide to Ancient Mexican Ruins* University of Oklahoma Press, 1977.

Irving, David. *The Mare's Nest* William Kimber, 1964.

Joyce, Thomas A. *Mexican Archaeology* Philip Lee Warner, 1914.

Kamil, Jill. *Luxor: A Guide to Ancient Thebes* Longman, 1973.

Lawrence, T.E. *Crusader Castles* Oxford, 1988.

Longmate, Norman. *Hitler's Rockets: The Story of the V-2s* Hutchinson, 1985.

Louis, Victor and Jennifer. *The Complete Guide to the Soviet Union* Michael Joseph, 1976.

Macadam, Alta. *Blue Guide: Rome and environs* A. & C. Black, 1989.

MacFarquhar, Roderick. *The Forbidden City* Reader's Digest, 1972.

Mackay, Donald A. *The Building of Manhattan* Harper & Row, 1987.

Masson, Georgina. *The Companion Guide to Rome* Collins, 1985.

Meyer, Karl E. *Teotihuacán* Reader's Digest, 1973.

Murname, William J. *The Penguin Guide to Ancient Egypt* Allen Lane, 1983.

Rice, B. Lewis. *Epigraphia Carnatica* Mysore Government Central Press, 1889.

Romer, John. *Romer's Egypt* Michael Joseph/Rainbird, 1982.

Ruffle, John. *Heritage of the Pharaohs* Phaidon, 1977.

Sanders, Catharine, Chris Stewart and Rhonda Evans. *The Rough Guide to China* Routledge and Kegan Paul, 1987.

Sandstrom, Gosta. *The Crossing of the Alps* Hutchinson, 1972.

Sandstrom, Gosta. *The History of Tunnelling* Barrie & Rockliffe, 1963.

Sivaramamurti, C. *Panorama of Jain Art* Times of India, 1983.

Smith, Rex Alan. *The Carving of Mount Rushmore* Abbeville Press, 1985.

Speer, Albert. *Inside the Third Reich* Weidenfeld and Nicolson, 1970.

Stierlin, Henri. *Ancient Mexican Architecture* Macdonald, 1968.

Stott, Carole. *Astronomy in Action* George Philip, 1989.

Thompson, J. Eric. *Mexico before Cortez* Charles Scribner's Sons, 1933.

Tompkins, Peter. *Mysteries of the Mexican Pyramids* Thames and Hudson, 1976.

Tupper, H. *To the Great Ocean* (Trans-Siberian Railway) Secker & Warburg, 1965.

von Hagen, Victor W. *The Aztec: Man and Tribe* New American Library, 1958.

Watkin, David. *A History of Western Architecture* Barrie & Jenkins, 1986.

Yarwood, Doreen. *The Architecture of Europe* Batsford, 1974.

Young, Richard. *The Flying Bomb* Ian Allan, 1978.

Zewen, Luo, Dick Wilson, J.P. Drege and H. Delahaye. *The Great Wall* Michael Joseph, 1981.

Index

Acknowledgements

Picture Credits

l = left; *r* = right; *c* = centre; *t* = top; *b* = bottom

13 Ancient Art and Architecture Collection; 14, 15 The Illustrated London News; 17 The Photo Source; 18 Topham Picture Source; 19*l*, *tr* and *cr* Marc Riboud/The John Hillelson Agency; 19*br* Bruce Coleman Inc; 20–23 Alex Webb/Magnum; 24 Robert Harding Picture Library; 25 Richard Laird/Susan Griggs Agency; 26 UPI/Bettmann; 27 The Image Bank; 28 UPI/Bettman; 29 The Image Bank; 30, 31 UPI/Bettmann; 33 Mark Wadlow/USSR Photo Library; 34 V. Shustov/Novosti; 35*tl* and *r* Tass; 35*b* USSR Photo Library; 38 Ancient Art and Architecture Collection; 39, 40 Robert Harding Picture Library; 41 John P. Stevens/Ancient Art and Architecture Collection; 42–3 David Hiser/Photographers Aspen; 44 Loren McIntyre; 45*l* Werner Forman Archive; 45*r* Hutchison Library; 46*tl* Robert Harding Picture Library; 46*tr* Ancient Art and Architecture Collection; 46*b* Tony Morrison/South American Pictures; 47*t* E. Streichan/The Photo Source; 47*b* Tony Morrison/South American Pictures; 49 Ancient Art and Architecture Collection; 50 Robert Harding Picture Library; 51 Topham Picture Source; 52, 53 The Image Bank; 54*t* Tony Stone Associates; 54*b* Stephanie Colasanti; 55 Robert Harding Picture Library; 56*t* The Image Bank; 56*b* Mischa Scorer/Hutchison Library; 57 Scala; 58–9 Marc Riboud/The John Hillelson Agency; 60 George Gerster/The John Hillelson Agency; 62, 63 Stephanie Colasanti; 65 Guildhall Library/Bridgeman Art Library; 66, 67 Ann Ronan Picture Library; 68*tl* Mary Evans Picture Library; 68*tr* The Mansell Collection; 69*tl* Sefton Photo Library; 69*tr* Mary Evans Picture Library; 69*b* Architectural Association; 70 Robert Harding Picture Library; 71 Ancient Art and Architecture Collection; 72 Topham Picture Source; 73, 74*tr* Robert Harding Picture Library; 74*b* The Image Bank; 75 Robert Harding Picture Library; 76 Stephanie Colasanti; 77 Robert Harding Picture Library; 78*c* Ann Ronan Picture Library; 78*bl* and *r* Hulton-Deutsch Collection; 78–9 Ann Ronan Picture Library; 79*r* Mary Evans Picture Library; 79*bl*, *c* and *r*, 80*tl* Hulton-Deutsch Collection; 80*tr* and *b*, 81 Roger-Viollet; 84 The Image Bank; 85 Angelo Hornak; 86, 87 UPI/Bettmann; 88 The Illustrated London News; 89 UPI/Bettmann; 90–1 Alan Smith/Tony Stone Associates; 92*t* Bruce Coleman Inc; 92*b* Rene Burri/Magnum; 93*t* Bruce Coleman Inc; 93*b* Art Seitz/Gamma Liaison/Frank Spooner Pictures; 95 Bavaria/Lauter; 96 Bavaria/Hans Schmied; 97*t* Bavaria/Martzik; 97*c* Susan Griggs Agency; 97*b* Bavaria/Holl; 98, 99 The Image Bank; 102 Popperfoto; 103*l* The Image Bank; 103*r* Robert Harding Picture Library; 105*t* Bruce Coleman Inc; 105*b* Mike Powell/All-Sport; 106*t* Andrea Pistolesi/The Image Bank; 106*b* The Image Bank; 109 Canadian National Tower; 110*t* Panda Associates Photography; 110*b* Canapress Photo Service; 112*l* Robert Harding Picture Library; 112*r* The Image Bank; 113*l* Orion; 113*r* Northern Picture Library; 114 S. Tucci/Gamma/Frank Spooner Pictures; 115 Mike Yamashita/Colorific!; 116*tl* and *r* Gamma/Frank Spooner Pictures; 116*b* Mike Yamashita/Colorific!; 117 Gamma/Frank Spooner Pictures; 118 Sophie Elbaz/Gamma/Frank Spooner Pictures; 119*t* Associated Press; 119*b* Associated Press/Topham Picture Source; 120*l* Gamma/Frank Spooner Pictures;; 123, 124, 125 Peter Menzel/Colorific!; 128 Marc Riboud/The John Hillelson Agency; 129 Georg Gerster/The John Hillelson Agency; 130 Sally and Richard Greenhill; 132, 133*t* Georg Gerster/The John Hillelson Agency; 133*b* Anthony J. Lambert; 134–5 Gilles Mermet/Gamma/Frank Spooner Pictures; 135*b* Mary Evans Picture Library; 136*t* Colin Jones/Impact Photos; 136*b* Marion Morrison/South American Pictures; 137 Bruce Coleman Inc; 138 The Illustrated London News; 139 UPI/Bettmann; 140*t* Robert Harding Picture Library; 140*b* The Mansell Collection; 141*l* Paul Slaughter/The Image Bank; 141*r* Hugh McKnight Photography; 142–3 Canadian Pacific Corporate Archives; 144 The Mansell Collection; 145*t* Anthony J. Lambert; 145*c* and *b* Canadian Pacific Corporate Archives;; 147*t* Philip Robinson/John Massey Stewart; 147*b* Mark Wadlow/USSR Photo Library; 148*t* Roger-Viollet; 148*b*, 149*tl* John Massey Stewart; 149*tr* Roger-Viollet; 149*b* Topham Picture Source; 150–1 Bruce Coleman Inc; 152, 153 UPI/Bettmann; 154*t* Leslie Garland; 154*b* The Image Bank; 155 Bruno Barbey/Magnum; 156–158, 159*t* and *cl* Gamma/Frank Spooner Pictures; 159*cr* and *b* ANP Foto; 160–1 Anthony J. Lambert; 161 Key Color; 162 Prisma; 163 Key Color; 164 Ancient Art and Architecture Collection; 165*l* Geoff Tompkinson/Aspect; 165*r* Lee E. Battaglia/Colorific!; 167 Robert Harding Picture Library; 168 Ancient Art and Architecture Collection; 169*l* Michael Holford; 169*r* Robert Harding Picture Library; 170–1 The Photo Source; 172*tl* James Austin; 172*br* Leslie Garland; 174 Robert Harding Picture Library; 175 The Photo Source; 177–179 Norwegian Contractors; 181–183 Charles Tait; 184–5 B. Clech/Sodel; 186 M. Brigand/Sodel; 187*tl* J. F. Le Cocguen/Sodel; 187*tr* and *b* P. Berenger/Sodel; 188 Goutier/Jerrican; 189, 190, 191*c* Adam Woolfitt/Susan Griggs Agency; 191*b* Goutier/Jerrican; 194–5 Scala; 196 Roger-Viollet; 197 Scala; 203 Ivazdi/Jerrican; 204, 205 CERN; 206–209 Seikan Corporation; 210*t* Roger-Viollet; 210*b* Hugh McKnight Photograph; 211 R. Kalyar/Magnum; 214 Science Photo Library; 214–216 Kermani/Gamma Liaison/Frank Spooner Pictures; 220*t* The Image Bank; 220*b* Science Photo Library; 221 The Image Bank; 222–3 Alexis Duclos/Gamma/Frank Spooner Pictures; 224, 225 Gamma/Frank Spooner Pictures.

Artwork Credits

Craig Austin: 27*r*.
Trevor Hill: 50–1, 66–7, 130–1, 162–3, 168–9, 179.
Andrew Popkiewicz: 31
Simon Roulstone: 15, 27*c* and *cb*, 41, 45, 55, 61, 73, 93, 100–1, 106–7, 111, 125, 204–5, 208–9, 220.
Paul Selvey: 108, 130*tl*, 136–7, 170, 172–3, 191, 216–17.
All maps by Technical Art Services Ltd.